THIS BOOK IS DEDICATED TO

CAPT SASWATA DAS

(1983-2009)

He laid down his life in the service of the country in the best tradition of the Army.

- Commanding Officer, 28 Inf. Div. Sig. Regt.

Rabbit Production Management

Kalyan Sundar Das

Nityananda Das

2014

Studium Press (India) Pvt. Ltd.

Rabbit Production Management

ISBN: 1-978-93-80012-64-3

Published by:

Studium Press (India) Pvt. Ltd.
4735/22, 2nd Floor, Prakash Deep Building (Near PNB),
Ansari Road, Darya Ganj, New Delhi-110002
Phone: +91-11-43240200-15 (15 lines); Fax: +91-11-43240215
jngovil@gmail.com; jngovil@hotmail.com

Digitally Printed at:

Replika Press Pvt. Ltd.

Preface

Considering to meet out the increase demand of meat for growing human population and the liking of companion animals, now-a-days rabbit farming has become one of the most popular and economic enterprise among other livestock species. Livestock owners are taking more interest in rabbit farming due to minimum initial investment, high reproductive ability and unique meat characteristics of this species. Therefore, it is very essential to know about the scientific rearing of rabbits for achieving optimum production and reproduction, and to run an economically viable livestock farm.

The whole text has been divided into seven chapters. Emphasis is given to four main pillars of Livestock Production Management *i.e.*, feeding, breeding, housing and health control. First chapter deals general information of rabbits along with some of the important systems and their role in the body. Second chapter describes briefly about the classification, place of origin and body characteristics of rabbits. Third chapter represents digestion procedure, feed resources, nutrient requirements and feeding different categories of rabbits. Fourth chapter deals with puberty, breeding, pregnancy, kindling, weaning, selection and breeding system of rabbit. Fifth chapter depicts basic requirements of rabbit housing, housing environment, constructional details, housing systems, and feeding and watering equipments. Sixth chapter briefly describes rabbit diseases caused by virus, bacteria, fungus, parasites, and miscellaneous factors along with their symptoms, treatment and preventive measures to control them. Seventh chapter deals with the way to calculate economics of rabbit farming. At the end of this manuscript, references have been cited for further reading.

We do not pretend to claim credit for any original contributions on the subject. This manuscript has been prepared on the basis of experience of authors and collection of materials from different scattered resources. Every endeavour has been made to represent the topics in a very simple language and a systematic manner so that the readers are able to grasp very easily.

The authors are grateful to the Head, Department of Livestock Production and Management, Chowdhury Sarwan Kumar Himachal Pradesh Krishi Viswavidyalaya, Palampur and Head, Department of Livestock Production and Management. Guru Angad Dev Veterinary and Animal Science University (GADVASU), Ludhiana for giving the opportunity to Dr. K. S. Das to work with Rabbit Farms. We also thank Sh. Muni Lal, Laboratory Attendant of Rabbit Farm, GADVASU, Ludhiana for his help, inspiration and taking photographs for this book.

We acknowledge with thanks our debt to all those individuals, writers, authors organizations and publishers whose books, reviews, reports, articles, journals, proceedings, magazines etc. that we have quoted in this book.

This book is intended to serve as a text book providing comprehensive basic and technical information to undergraduate and postgraduate students of Veterinary and Animal Sciences, teachers, researchers, academicians, veterinarians, stock persons or farm managers and pet lovers who are engaged and interested in this field. We welcome suggestions, criticisms and comments from the readers for future improvement.

The co-operation and help of M/S Studium Press (India) Pvt. Ltd., New Delhi to bring out this book is highly appreciated.

We are indebted to our octogenarian parents for their cornucopian love, affection and blessings, and our elders, youngers, son and daughters for their encouragement in preparation of this manuscript.

Finally, we wish to put our appreciation and gratitude to Mrs. Bandana Das, wife of Dr. K. S. Das for her contribution in artistic drawing and preparation of different diagrams for this book, her constant help and support in writing, constructive suggestions and encouragement. We also owe the same debt of gratitude to Mrs. Chandana Das, wife of Dr. N. Das for her inspiration, unfathomable perseverance and constant help and support in writing of this book.

Kalyan Sundar Das

Nityananda Das

About the Authors

Dr. Kalyan Sundar Das

Dr. Kalyan Sundar Das, Son of Sh. Pulin Bihari Das, has born on 24th July, 1972 at Bhemua, Paschim Medinipur, West Bengal. He passed his B.V. Sc. & A.H. from Bidhan Chandra Krishi Vishwavidyalaya, West Bengal in 1995. He did both his Masters of Veterinary Sciences and Doctor of Philosophy in Livestock Production and Management from Indian Veterinary Research Institute Izatnagar, Bareilly (U.P.). He worked as Assistant Professor in the CSK Himachal Pradesh Krishi Viswavidyalaya, Palampur (H.P.) and the Punjab Agricultural University, Ludhiana. Presently, he is working as Senior Scientist at the Regional Station of Central Institute for Research on Buffaloes, Nabha, Patiala (Punjab). He has more than 13 years of experience in the field of research, teaching and extension. Besides 38 research papers of national and international repute, he has published four laboratory manuals and eight farmers' training manuals.

Dr. Nityananda Das

Dr. N. Das is currently working as Principal Scientist at Indian Grassland and Fodder Research Institute, Jhansi (UP). He has 28 years of research experience in the field of Livestock Production and Management. As deputy advisor Planning Commission, Government of India, he was associated with the identification of policies and strategies for development of animal husbandry sector in India. He was also associated with the Mid-term appraisal of the Ninth (1997–2002) and formulation of the Tenth Five Year Plan (2002–2007) programme in the livestock sector. He is one of the pioneer workers in the field of animal behaviour. He has guided many postgraduate and PhD students at IVRI, Izatnagar. Dr. Das has 50-research publications of national and international repute. He also edited two books 'Forage for Sustainable Livestock Production' and 'Nutrition and Care of Livestock during Natural Disaster'.

Contents

Chapter 1

General Considerations of Rabbits

ORIGIN AND DOMESTICATION OF RABBIT

Rabbits belong to the order of mammals called Lagomorpha that includes 40 or so species of rabbits, hares and pikas. Lagomorpha evolved in Asia at least 40 million years ago, during the Eocene period. The family Leporidae (rabbits and hares) has been relatively unchanged since the Eocene Epoch about 40 million years ago. Rabbits had entered North America by that time, and they underwent most of their development there. By about seven million years ago (the Miocene Epoch), they had become re-established in Asia and had moved into Europe, which led to the present distribution.

The wild rabbit *Oryctolagus cuniculus* of southern Europe and North Africa was associated with the human being over 1000 years B.C. and was first recorded by the Phoenicians, when they termed the Iberian peninsula '*i-shephan-im*' (literally, 'the land of the rabbit'), which the Romans converted to the Latin form, Hispania, hence the modern word Spain. The wild rabbit has long been hunted, but, it is unclear exactly when first domestication took place. It is believed that rabbits are originated from Spain. In Roman times the rabbit was symbolic of Spain. The Romans kept rabbit in walled gardens with deep foundations to the walls so that they could not burrow out. Rabbit enclosure which was known as ‘Coneygarth’ consisted of a small field surrounded by trees and hedgerow. It appears that the Romans spread the rabbit throughout the Roman Empire as a game animal.

In Europe, and especially France, the domestication process was well under way by the fifth century, and in the twelfth century the Normans brought them to Britain, where they became established and remained as both domestic and wild animal. Credit for the actual domestication of this species goes to the early French Catholic monks who wanted a food suitable for Lent. Unborn or newly born rabbits, known as ‘laurices’ (officially classified as fish in 600 A.D. by Pope Gregory I), was much favoured by the Roman. Monks were in the habit of eating *laurices* during Lent as they were considered “an aquatic dish”. Although breeding experiment started in French monasteries from the 6th century, it was not successful until 16th century. Several breeds of rabbit (grey-brown wild, white, black, piebald and ash-grey rabbits) were known in the sixteenth century and this was the first indication of controlled breeding. In 1606, Olivier de Serres classified three types of rabbit: the wild rabbit, the semi-wild or “warren” rabbit raised inside walls or ditches, and the domesticated or hutch-bred rabbit. Domestic breeding began during 17th century. It was mainly the work of monks, since it provided them a delicious dish compared to the tougher wild rabbit. Later, Emperor Napoleon III of France established rabbit breeding in smallholding scale. This produces millions of rabbit pelts that were used to line solder’s cloaks.

Rabbit was not accepted as popular animal until 19th century. Commercial rabbit farming started in the last century. The scarcity of food following First and Second World War pave the way for extensive rabbit production. It is considered as ‘underground mutton’ due to the excellent meat quality. Food and Agricultural Organization prepared various package of programmes for farmers to boost rabbit farming. Then, it spreads to other countries in the world. In developed countries of Western Europe *e.g.*, France, Italy, Spain, Hungary, U.K., Germany, Denmark, U.S.A. etc. rabbit industry has been established to its highest degree and become an essential part of livestock enterprise. Among developed countries, China has emerged as a leading country and is a major producer of rabbit meat, fur and Angora wool in the world.

In India, no systematic efforts were made to uplift this industry by either government sector or private sector or breeders association or other organizations before 1960's. A few years after that, during 1978–79, Indian Council of Agricultural Research (ICAR) imported rabbits from the U.K. and former U.S.S.R. for meat and wool production and took many steps to establish and popularise this species for the Indian farmers. The Council launched various research and development projects on rabbit production at the Division of Fur Animal Breeding, Garsa, Himachal Pradesh which is a regional station of Central Sheep and Wool Research Institute (CSWRI), Avikanagar. The research technologies developed at the stations are being transferred through various extension activities, training programmes and publications. Now-a-days, the rabbit farming is practiced throughout the temperate region of our country *viz.*, Himachal Pradesh, Jammu & Kashmir, Uttarakhand, Darjeeling, Sikkim, Shillong and even, in Ooty of Tamil Nadu. A number of good rabbit strains suitable under Indian climatic conditions have been developed by the scientists of North Temperate Regional Station (NTRS) of CSWRI. Defence Agriculture Research Laboratory, Pithoragarh has also played very important role to popularize this farming in Munsiyari, Darkot and Pithoragarh city in Uttarakhand. As a result, rabbit is not only commercially raised as meat, skin and wool producing animal but also used as laboratory, pet or companion animal throughout the country.

TAXONOMY

Rabbit does not belong to one single taxonomic group. Rather, the name refers to an accumulation of 10 genera in the family Leporidae whose characteristics are intermediate between hares and pikas, the other members of order Lagomorpha. Out of 10 genera in the family Leporidae, two genera *Pronolagus* and *Caprolagus* are considered as hare.

All the domestic rabbits are considered to be the descendent of European wild rabbit belonging to hares (Genus-*Lepus*) and American Cotton Tail rabbits (Genus-*Sylvilagus*).

Kingdom - Animalia

Phyllum - Chordata (Animal with notochords and gills)

Sub-phyllum - Vertebrata (Chordates with organized head region)

Class - Mammalia (Warm blooded craniates with hair coat and young nourished from mammary gland)

Sub-class - Theria (Viviparous mammals)

Order - Lagomorpha (Characterised by gnawing placenta with chisel like incisors of which there is in the upper jaw a small second pair directly behind the main pair and coitus induced ovulation)

Family - Leporidae (Rabbits and hares)

Sub-family - Leporinae

Genus - There are total 8 genera with 24 species

1. ***Sylvilagus (Cottontail rabbit)*:** Sixteen species of North and South America.
    - Forest Rabbit, *Sylvilagus brasiliensis,*

- Dice's Cottontail, *Sylvilagus dicei*
- Brush Rabbit, *Sylvilagus bachmani*
- San Jose Brush Rabbit, *Sylvilagus mansuetus*
- Swamp Rabbit, *Sylvilagus aquaticus*
- Marsh Rabbit, *Sylvilagus palustris*
- Eastern Cottontail, *Sylvilagus floridanus*
- New England Cottontail, *Sylvilagus transitionalis*
- Mountain Cottontail, *Sylvilagus nuttallii*
- Desert Cottontail, *Sylvilagus audubonii*
- Omilteme Cottontail, *Sylvilagus insonus*
- Mexican Cottontail, *Sylvilagus cunicularis*
- Tres Marias Rabbit, *Sylvilagus graysoni*
- Manzano Mountain Cottontail, *Sylvilagus cognatus*
- Appalachian Cottontail, *Sylvilagus obscurus*
- Robust Cottontail, *Sylvilagus robustus*

2. ***Nesolagus (Striped rabbits)*:** Two Southeast Asian species.
   - Sumatran Striped Rabbit, *Nesolagus netscheri*, and
   - Annamite Striped Rabbit, *Nesolagus timminsi*
3. ***Brachylagus* (*Pygmy rabbit*):** One species of the United States
   - *Brachylagus idahoensis*
4. ***Bunolagus* (*Riverine rabbit*):** One African species.
   - Bushman Rabbit, *Bunolagus monticularis*
5. ***Oryctolagus* (*European rabbit*):** One European Rabbit, *Oryctolagus cuniculus*, introduced to Australia and numerous other countries.
6. ***Pentalagus* (*Amami rabbit*):** One Japanese species.
   - *Pentalagus furnessi*

7. ***Poelagus* (*Bunyoro rabbit*):** One African species.
    - Central African Rabbit, *Poelagus marjorita*
8. ***Romerolagus* (Volcano rabbit, *zacatuche*):** One Mexican species.
    - *Romerolagus diazi*

The best-known and most recognizable are the European rabbit which includes 16 species of North and South American cottontails. The European rabbit was originally found in the Iberian Peninsula and Northwestern Africa, but, subsequently it dispersed throughout Western Europe 2,000 years ago. European rabbits are also introduced in oceanic islands throughout the world, parts of Chile and Argentina, and also New Zealand and Australia, where it thrives. Most cottontails prefer open or brushy habitats, although some live in tropical forests and others are semiaquatic (the swamp rabbit, *S. aquaticus*, and the marsh rabbit, *S. palustris*). There are five rabbit species live in Africa. The Bunyoro rabbit (*Poelagus majorita*) is found in Central Africa, while the three species of Rockhares (genus *Pronolagus*) are found in Southern Africa and they inhabit rocky areas associated with grass or woodlands. The riverine African rabbit (*Bunolagus monticularis*) is available in Karoo region of South Africa, where it inhabits dense vegetation along seasonal rivers and it becomes a endangered species because of gradual habitat destruction throughout its home track. The Asia rabbit (hispid hare, *Caprolagus hispidus*) occupies the dense, tall grassland (elephant grass) along the southern Himalayan foothills of Nepal, Bangladesh, and India and like the riverine African rabbit. The Asian rabbit is also endangered. The Amami rabbit (*Pentalagus*) lives only in forests on two small islands (Amami and Tokunoshima) of Japan and it is also an endangered species. The Sumatran rabbit (*Nesolagus netscheri*) lives in the island's southwestern montane forests and it is in the verge of extinction. Although there has been only one confirmed sighting of this animal since 1916, two pictures of the rabbit were captured with an automated camera in 1998. Three African species under genus *Pronolagus* (Rock hares) and one Indian (Asian) species of Caprolagus genus (Hispid hare) are commonly known as rabbit but they are actually hare.

DIFFERENCE BETWEEN HARE AND RABBIT

Hares and Jackrabbits are leporids belonging to the genus *Lepus*. Four species commonly known as types of hare are classified outside of *Lepus*: the hispid hare (*Caprolagus hispidus*), and three species known as red rock hares (*Pronolagus* species). Like other leporids, hares do not rear their young below ground in a burrow, but rather in a shallow depression or flattened nest of grass called a form. The basic difference between rabbits and hares is how their kits are born. Hares are precocial and hence soon after birth they are able to fend for themselves. Young hares, being born fully furred and with eyes open, do not require much physical protection. On the other hand, rabbits are altricial, having young that are born blind and hairless. Morphologically, rabbits and hares have a more arched skull correlated with development of bounding locomotion and a relatively upright posture of the head. Strengthened hind limbs and pelvic girdle and elongation of the limbs are also evident.

Hare	***Rabbit***
1. They have larger body size, longer hind leg and ears, and have black marking on their fur.	1. They have smaller body size, comparatively shorter hind leg and ears.

Table: (*Contd...*)

Fig. 1: Cape hare

Table: (*Contd...*)

Hare	*Rabbit*
2. They are born with full coat of fur and eyes open. Mother drops them on bare ground at birth or into a slight depression in the ground.	2. They are born hairless and eyes close. Mother prepares a nest with grass, bark and soft stems and over this, she places a layer of hair that she plucks from her own body. When mother leaves nest, she covers young with more hair and dead plants to keep them warm and hidden from enemies.
3. Fur of hare in cold snowy region turns white in the winter.	3. In rabbit, brown summer fur is replaced with fur that is greyer.
4. Hares live in simple nests above the ground. They hides among plants and usually try to escape enemies by running.	4. All rabbit (except the cotton-tail rabbits) hides in either burrows or depressions in the ground during day.
5. Hare lives most of the time by themselves. They come	5. Rabbits are often found together. To become dominant

Table: (*Contd...*)

Table: *(Contd...)*

Hare	*Rabbit*
together in pairs for mating only. There is little or no fighting among hares.	male, males fight within group. The dominant male rabbits only mate with most of the females in the area.
6. Hares are not domesticated.	6. European rabbits are domesticated and they are raised for meat and also kept as pets.

Fig. 2: European hare

Fig. 3: Jack rabbit

RELATED TERMINOLOGIES

Terminology	*Description*
Buck	A mature male rabbit used for breeding
Bunny	A young rabbit below twenty weeks of age
Doe	A mature female rabbit used for breeding
Doe in kindle	A female rabbit carrying young
Fur	Wool from rabbit
Fryer	A ten to twelve weeks old rabbit ready for market
Kindling	Act of parturition in rabbit
Kitten/kit	A young rabbit whose eyes are not yet opened
Litter	Kits born in a single kindling or, bedding materials used for rearing rabbit
Pelt	Skin of rabbit
Rabbitry	The husbandry of rabbits
Roaster	A culled rabbit
Weaner	A newly weaned rabbit

WORLD RABBIT PRODUCTION

As per report of Food and Agricultural Organization (2010), recent rabbit population in the world comes around 828 millions. The rabbit population scenario is changing in the global perspective. Now-a-days, it is becoming more popular in Asia (67 to 68% of world population) compared to other countries in the world. The credit of lion share of world rabbit population goes to China (Table 1). The share of American and European countries in terms of population is almost similar in the international level. Number of slaughtered rabbits, average carcass weight and meat production of major rabbit producing countries of world are given below.

Rabbit Population in India

As per the report of Anon (2004), the total rabbit population in India is 4.06 lakhs. The major rabbit populated states of

Table 1: Major rabbit producing countries and their production

Country	*Rabbit and hare stock (1000 head)*	*Rabbit meat producing animals/ slaughtered (1000 head)*	*Yield / carcass weight (gm/ animal)*	*Production quantity (tonnes)*
Austria	300	200	1700	340
Belarus	280	400	1750	700
Belgium	174	NA	NA	NA
Brazil	226	1100	1500	1650
China	290040	514600	1300	669000
Czech Republic	5500	20000	1925	38500
Egypt	7300	45500	1200	54600
France	10900	36777	1405	51665
Germany	13000	22000	1705	37500
Hungary	815	3946	1370	5404
Italy	72000	45500	1505	255400
Mexico	1350	4250	1059	4500
Poland	632	1234	2350	2900
Romania	1700	135	1215	164
Spain	3200	58800	1126	66200
Ukraine	5621	7314	1846	13500
Africa (Total)	12471	69035	1165	80423
America (Total)	130191	149531	1855	277394
Asia (Total)	561353	621363	1302	809071
Europe (Total)	123876	340712	1494	509107
World (Total)	827891	1180641	1420	1675995

NA= Not available; *Source:* FAO Production Year Book (2010)

India are Kerala (1.48223 lakhs), Nagaland (0.38408 lakhs), Bihar (0.29380 lakhs), Rajasthan (0.22929 lakhs) and Uttar Pradesh (0.22191 lakhs). The approximate number of Angora rabbit is 0.48 to 0.69 lakhs (Singh *et al.*, 2004).

ADVANTAGES OF RABBIT FARMING

- Backyard rabbitries are particularly well suited to small, marginal and landless farmers as they are highly productive in terms of offspring (kg/year/dam). Shorter generation interval and high prolificacy give quick return with low initial investment in farming.
- The main emphasis of meat production is to convert plant proteins of little or no use to people as food into high-value animal protein (meat). As a small monogastric herbivore, the rabbit can thrive on a fairly wide range of cellulose-rich foods. They can utilize kitchen and garden refuse, weeds, plants, leaves etc. and thus, reduces cost of production.
- They are good converter of low quality plant protein to high quality animal protein in terms of meat. Rabbits can turn 20 percent of the protein they eat into edible meat compared to 22 to 23 percent for broiler chickens, 16 to 18 percent for pigs and 8 to 12 percent for beef. In compare to chicken, rabbits can more efficiently convert the available proteins in cellulose-rich plants. The traditional grain fed to the poultry put them in direct competition with humans for food. For countries with no cereal surpluses, rabbit meat production is thus advantageous.
- Rabbit is prolific breeder (Table 2) and has fast growth rate. When cattle, goat or sheep are raised for meat production, most of the energy consumed by the herd or flock is used to maintain breeding females which have a low prolificacy: a maximum of 0.8 to 3.0 young per year against 40 for female rabbits.
- Rabbitries is a flexible business which ranges from backyard rabbit keeping to large commercial farms.Investment in infrastructure and equipment is low and not much space is needed.

- Rabbit is adaptable to different feeding and holding conditions. It is adaptable to the family diet and food preservation techniques available on small rural and peri-urban farms.
- Labour costs are low and the work can be done by family members: women and children, or perhaps aged or handicapped people, for whom rabbit husbandry, like that of other small animals, represents an attractive and remunerative occupation.
- Rabbit meat is low in fat and cholesterol (Table 2), and rich in protein (Table 3) and sodium which can be recommended for cardiac patients.
- The heat insulation of Angora wool fibre is 2–3 times higher than that of sheep wool fibre in relation to weight.
- To produce one kg pure Angora wool, Angora rabbit need only 30% of the digestible energy required by sheep and 90% of that required by Angora goats.
- Meat rabbits attain their market age very quickly.
- The energy expenditure per kg weight gain is 39% less than sheep and 47% less than cattle.
- Angora wool and rabbit meat have good market value.
- There is no religious taboo for meat consumption among non-vegetarian community throughout the country.
- The skin of broiler rabbit can be used for making caps, gloves, jackets, blankets, ladies purses and attractive decorative items.
- Rabbit faeces, being rich in protein, serve as good manure and can be used in animal feed after proper processing up-to 10% without affecting production.
- Rabbit faeces are a very good material for growing composting worm (*Elsemis foetida*).

Table 2: Performance of different animal species

Livestock species	***No. of young per breeding female per year***	***Live-weight of breeding female***	***Live-weight at slaughter***	***Slaughter yield***	***Daily weight increase***		***Fat content of carcass***	***Food kcal per g of usable protein***
		kg	***kg***	*%*	***g***	***g/kgW***$^{0.75}$	*%*	***kcal/g***
Broilers	100	3.0	2.0	63	31	30.5	13.0	80
Turkeys	60	10	10.1	79	65	19.2	13.0	87
Rabbits	40	4.5	3.2	60	32	22.3	6.8	105
Pigs	12	170	100	73	540	28.4	32.0	151
Sheep	1.4	70	50	50	220	18.2	36.0	427
Beef cattle	0.8	500	475	61	950	14.8	32.0	442
Dairy cattle[1]	0.8	500	475	61	950	14.8	32.0	184

Source: Dickerson (1978)

Table 3: Meat composition of different livestock species (per 100 g meat)

Type of meat	*Energy*	*Water*	*Crude proteins*	*Crude fats*	*Crude ash*	*Calcium*	*Phosphorus*	*Potassium*	*Sodium*	*Iron*	*Vitamins*					
											A	B_1	B_2	B_6	*Nicotinic*	*Calcium pantothenate*
	kcal	*g*				*mg*					*UI*	*mg*				
Chicken mutton	200	67	19.5	12	1	10	240	300	70	1.5	200	0.05	0.10	0.45	8	0.90
Lean meat	210	66	18	14.5	1.4	10	165	350	75	1.5	40	0.15	0.20	0.3	5	0.55
Fatty meat	345	53	15	31	1	10	130	350	75	1	80	0.15	0.20	0.3	4.5	0.55
Pork																
Lean meat	260	61	17	21	0.8	10	195	350	70	2.5	tr	0.85	0.20	0.3	4.5	0.50
Fatty meat	330	54.5	15	29.5	0.6	9	170	350	70	2.2	tr	0.70	0.15	0.3	4	0.50
Rabbit beef	160	70	21	8	1	20	350	300	40	1.5	-	0.10	0.05	0.45	13	0.80
Lean meat	195	66.5	20	12	1	12	195	350	65	3	40	0.10	0.20	1.5	5	0.45
Fatty meat	380	49	15.5	35	0.7	8	140	350	65	2.5	90	0.05	0.15	1.5	4	0.45

Source: Adrian *et al.* (1981)

CONSTRAINTS OF RABBIT FARMING

In spite of excellent productive capability of rabbits, this enterprise has not become a principal component of animal agriculture in our country due to following reasons:

- Rabbits are not adaptable to fluctuating environment / climatic conditions.
- Lack of superior germplasm.
- Lack of expertise in providing scientific management practices for rabbit rearing.
- Cost of concentrate ingredients is high and competes with the food grains of human being.
- Lack of marketing infrastructure for disposal of rabbit products.

MORPHOLOGY OF RABBIT

The body parts of rabbit can be divided into four main parts *viz.*, head, neck, trunk, and tail.

1. ***Head*:** The head is ovoid, somewhat narrow in front and it makes an angle with the rest of the body. It bears mouth, external nares or nostrils, eyes, ears and vibrissae.

   A. **Mouth:** The mouth lies at the lower anterior end of the head. It is relatively small, transverse slit and is bounded by two movable lips-upper and lower. The upper lip is larger and divided by median vertical cleft into right and left halves which is known as hare lip. The halves of the hare lip can be retracted to expose the front incisor teeth. The cleft connects the mouth with the nares above.

   B. **Nostrils or external nares:** The external nares are oblique and slits. They lie under the tip of the snout. Within each nostril there is an oval sensory pad. The nostrils are surrounded by a small, moist and bare area of skin, called *rhinarium*.

C. **Eyes:** A pair of prominent eyes is located from the some distance above the external nares. They are laterally placed near the upper side of the head. Each eye has three eyelids-upper, lower and third eyelid called nictitating membrane that moves across the cornea and offers protection and lubrication. The upper and lower eyelids are movable and bear a few eyelashes. The eyes have a field of view of 360°. Practically, the fields of view of the right and left eye overlap by about 30° in front and 9° behind. It provides them panoramic field of vision in order to better view predators. Rabbit possesses a double retinal system of rods and cones in the eye resulting eight times greater light sensitivity than humans. Surprisingly, their eyes readily detect motion and apparently are sensitive to the blue and green present at twilight. They are unable to view the small area below the mouth, so they depend on the sensitive lips and whiskers to aid them in finding food.

D. **Ears:** A pair of long movable, trumpet-shaped flaps, pinnae, is present behind and above the eyes. Each pinna leads into a narrow tube, the external auditory canal, which is closed internally by the ear drum or tympanum. The external auditory canal and the pinna together form the external ear. The pinnae are held upright when the rabbit is on the alert, but, are laid back when it is frightened or running. It collects sound wave and enable the rabbit to know accurately the direction from which a sound is coming. The tips of pinnae are sensitive to touch and help the rabbit to judge the height of the burrow. It comprises about 12% of the total body surface. Rabbit ears contain an abundance of blood vessels. The function of ears is just like a 'car radiator' which helps to dissipate heat during summer months.

E. **Vibrissae:** Large number of long, stiff, sensory hairs, called vibrissae whiskers, is present on both sides of the mouth, nares, above and below the eyes. They act as organs of touch and are very helpful in the dark burrow. The length of whiskers of the two sides taken together is more than the width of the body. They enable the rabbit to know whether it would be able to pass through a passage or not.

2. ***Neck:*** Neck is quite short in length and connects the head with the trunk. It permits the animal to turn its head to some extent. This enables the rabbit to direct the special sense organs to the place to which its attention is drawn. The neck looks thicker than it really is due to the loose skin and dense fur which protect this very delicate region.

3. ***Trunk:*** The shape of trunk is somewhat cylindrical with three prominent regions-vertebral column, thorax and abdomen.

   A. **Vertebral column:** The portion extended from the upper part of shoulder region to the joint of last rib in the spinal cord is called back. The part from back to the starting of hip bone is loin which also contains loose skin. The distance between loin and the base of the tail is known as rump. It is round in shape.

   B. **Thorax:** This is the small and narrow anterior region. It is firm ventrally as it is enclosed by ribs and sternum and contains heart and lungs.

   C. **Abdomen:** This is the large broad posterior region. It is soft being enclosed by body wall only and contains viscera. In female rabbit, on the ventral surface of the thorax and abdomen, there are four pairs of projections which are called teats or nipples. In male rabbit, teats are also present, but, in vestigial condition.

(*i*) *Apertures*: There are two apertures at the end of the trunk.

(*a*) *Anus*: This is an outlet for the alimentary canal. It lies below the base of the tail.

(*b*) *Urogenital aperture*: It lies in front of the trunk. In male rabbit, it is situated at the tip of a cylindrical structure, called the penis. The tip of the penis is very sensitive and is termed as glans penis which is enclosed in a fold of loose skin, the prepuce. In the either side of penis, there is an oval pouch of skin, scrotal sac, containing a testis. The urogenital aperture in female rabbit is slit like and is known as vulva. Within the vulva at its anterior margin, there is a small rod like body, the clitoris which is a homologous to the penis of male.

(*ii*) *Perineal pouch*: A pair of hairless depressions, lie on either sides but a little in front of anus in both the sexes, is called perineal pouch. Each pouch contains papilla on which perineal gland opens. The secretion of perineal glands gives a characteristic odour to the rabbit.

(*iii*) *Limbs*: The trunk bears one pair of fore limbs at the anterior and one pair of hind limbs at the posterior end ventrolaterally. They are vertical in position to hold the body clear of the ground, when required.

(*a*) *Fore limbs:* The fore limbs are shorter than the hind limbs. Each for limb consists of upper arm of brachium directed backwards and downwards from the body, fore arm or antebrachium directed forwards and downwards from the elbow and hand or manus, directed forward. The upper arm is largely concealed by the skin and fur of the body wall. The hand further consists of the wrist or carpus, the palm or metacarpus and five fingers or digits. Each finger ends in a sharp, curved and horny claw. The first or inner finger is the shortest and is called thumb or pollex. The third finger is longest. The lower surface of the hand is hairy and is known as palm. The fore limbs act as chief digging organs.

(b) *Hind limbs*: The hind limbs are longer and more powerful than fore limbs. They act as chief locomotory organs. Each limb is divided into the proximal thigh or femur directed forwards, the middle shank or crus directed backwards and downwards from the knee, and the distal foot or pes directed forwards. The foot is composed of the ankle or tarsus, the instep or metatarsus and four toes or digits. The first digit or hallux is absent. Each toe ends in a sharp, curved and horny claw. The lower surface is covered by hair and is known as sole.

4. ***Tail:*** The tail is the posterior end of trunk and above of the anus. The tail is composed of the coccygeal vertebrae, tendons, ligaments, arteries, veins, nerves, muscles and skin. The last bone of tail is pointed. It is short, bushy, and movable, generally curved upwards and used to give the danger signal to other rabbits. Tail flagging may occur during the breeding ritual.

Rabbits have three pairs of scent glands present in chin, anus and inguinal regions by which they mark their territories. The size of gland and marking behaviour are associated with the level of sexual activity of rabbit. The bucks generally mark more than does. On the other hand, does use these glands to mark their own babies. Due to the marking of territorial behaviour, at the time of hand mating of rabbit, the doe is taken to buck not vice versa. If male is taken to female territory, the female dominates in the majority of cases and does not allow mating the male in her territory resulting unsuccessful mating.

HANDLING OF RABBITS

Domestic rabbits are very docile and can easily be handled even by the children. Soft spoken words with hand pat, relax, calm and quiet approach are the pre-requisites of rabbit handling. Confidence of handler is also essential for this job. Frightened and mishandled rabbits can kick forcefully with their hind legs and scratch with their claws, or bite. Proper handling is beneficial both for the rabbit's and the handler's well-being. Moreover, they have light weight skeleton compared to most animals and their bones are easily fractured. If they struggle violently, the lightning-fast movements of the rear legs cause over-extension of the lumbosacral region of spine which frequently results in fracture or dislocation or sometimes die of heart failure. The handler should never try to overpower a struggling rabbit. When the rabbit violently resists the physical restraint it should be immediately released and approached after becoming calm down. Restrain the rabbit placing one hand of handler in front of head of animal and other hand on the back side. The rabbits should never be picked up by their ears or by their hind legs. The following methods can be adopted for handling of different categories of rabbits.

- The smaller and young rabbits are lifted in one hand through holding the loose skin either of neck region or of loin area of rabbit (Fig. 4). This method is not practiced in pregnant rabbit because it gives pressure on gravid uterus and may cause abortion.

- The correct method of handling a medium and pregnant rabbit is to grasp the loose skin over the neck with the help of one palm. Pass the fore-finger through the area in between two bases of ears and encircle the bases of ears with the help of middle finger and thumb. Lift the rabbit after giving support the rump of rabbit with other hand (Fig. 5). Remember that uncomfortable rabbit will try to struggle and to scratch the handler.

Fig. 4: Correct way of handling of small and young rabbits

Fig. 5: Correct way of handling of medium and pregnant rabbits

- Large or heavy rabbit breeds are handled like medium size breeds. In this case, the rabbit is placed in between hand and body of handler to the either side. The legs of rabbit should be placed away from and back side of the handler's body to avoid scratching to the handler.
- In docile rabbit, hold the rabbit in left hand as in handling of medium rabbit and right hand is set parallel to the handler's body. Put the rabbit on the right hand, release the neck and pat with left hand (Fig. 6).

Fig. 6: Correct way of handling docile rabbit

- To handle furious and frightened rabbits the method followed is similar as in handling of docile rabbits. But, the eyes and head of rabbits are placed in between the elbow and body of handler so that the rabbits can't see the surrounding environment (Fig. 7).
- The rabbits can also be wrapped with towel to handle them for collecting blood from ear vein or for oral examination. To do this, first place the rabbit on top of a flat towel and then, wrap towel tightly around legs and body starting from shoulder down to the hips.
- Rabbits can easily enter in the torpid (hypnotised) state when placed and held on their backs for few seconds. The effect is pronounced if they are blindfolded.

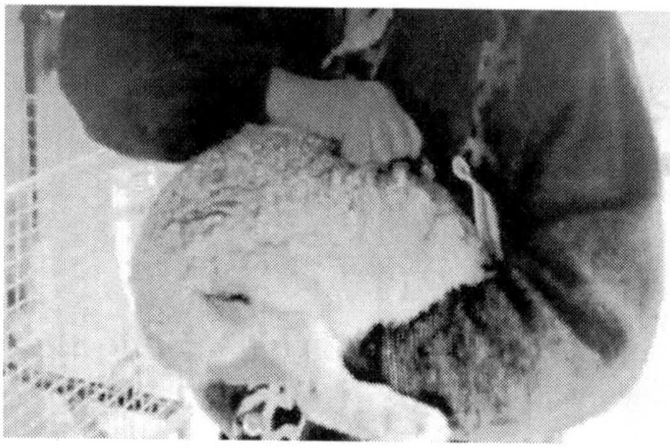

Fig. 7: Correct way of handling furious or frightened rabbits

- The furious rabbits can be handled by using sedative drugs like Ketamine-hydrochloride @ 20–44 mg/kg body weight, Tiletamine-zolazepam @ 10–15 mg/kg body weight and Fentanyl-droperidol @ 0.15–0.17 mg/kg body weight.

HANDLING OF RABBITS DURING TRANSPORT

Rabbits are very sensitive to handle. When they are handled to transport from one place to another by transport vehicle utmost care is must to reduce transport stress.

- Use separate container or crate or basket with a lid that can be closed for each rabbit to transport them. There should be sufficient space for ventilation.
- Move rabbits very quietly and gently. Care must be taken not to frighten them. They may get sick or die.
- Handle them during comfortable part of the day. It should be done in the morning during summer and noon during winter.
- Allow only experienced persons to handle them.
- Load them carefully in the transport vehicle.

- Do not move rabbits with full stomach. It is better not to offer any food a night before to transport them for the short journey. Rather, put some food and water in the container to feel them at home to settle down quickly. To transport them for long distance give sufficient rest at every 6–8 hours journey.
- Protect them from the bad climate in the journey. If possible, use air conditioned vehicle to minimize stress.
- Unload the container with care after reaching the destination.

DIGESTIVE SYSTEM

Rabbits are true non-ruminant herbivores. Their digestive reservoir permits and increases the efficiency of utilization of fibrous diets. They have a large stomach and well-developed ceacum relative to other non-ruminant herbivores such as the horse. The high voluntary feed intake (VFI) is at least 4 times higher pro rata than a 250 kg cattle. It is also associated with a low gut retention time of 17 hours in the rabbit compared with 68 hours in the bovine. High VFI together with re-utilization of gut content by reingestion of cecal material supports the rabbit's high nutrient requirement per unit of body weight and improves feed utilization for the rabbit. The digestive system of rabbit is shown in Fig. 8.

The rabbit has a large mandible with small oral cavity. The teeth of rabbit are 'open rooted' or 'aradicular hypsodont' and continuously growing in nature. The deciduous teeth are non-functional and shed perinatal. The permanent teeth are completely erupted by two to five weeks of age. The teeth consist of two pairs of incisors or 'çutting teeth' on the upper jaw and one pair on the lower, three pre-molars and three molars on either side of the upper jaw, and two pre-molars and three molars on either side of the lower jaw. Thus, the dental formula for a rabbit is: 2 x (I 2/1: C 0/0: P

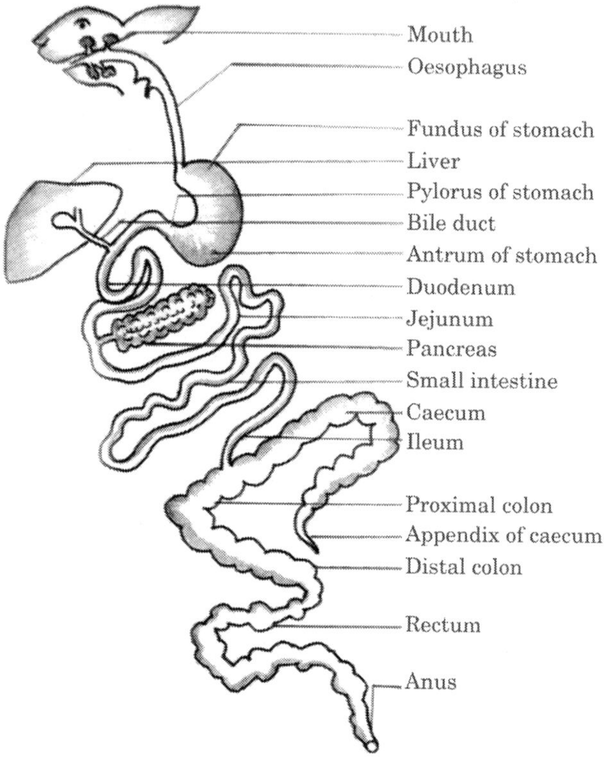

Fig. 8: Digestive system of rabbit

3/2: M 3/3) = 28. The second pair of upper incisors or "peg teeth" is normally small and present behind the upper incisors and may be difficult to see. The second pair of incisors may be absent in some rabbits. The lower incisors fit between the large upper incisor and peg teeth. The upper incisors erupt approximately 2 mm per week which is slower than lower incisors, which grow at the rate of 2.4 mm per week. Dental growth depends on the age, nutrition, type of feeds, wearing of teeth, pregnancy, health of animals etc. The chisel like incisors is used for lateral slicing which cut grasses before being chewed by the cheek teeth. The premolar and molar teeth are known as cheek teeth. The cheek teeth, situates in the back of the mouth, are arranged in parallel rows with flat chewing surface and prominent ridges. As the size of last molar tooth of upper jaw is small, five pairs of grinding teeth are functional on each side. The jaw can move up and down, forward and back, side to side for chewing.

The rabbit feeds frequently–20 to 30 times per day of 2–8 g of food over 4–6 minute periods. Mouth of rabbit is connected by several well developed duct from the salivary gland. These glands secrete saliva to digest food materials. From the mouth, a slender tube, the oesophagus, carries food into the stomach. Oesophagus is one way passage which prevents the ability to vomit or cough hair balls.

The stomach of the rabbit, which is J-shaped, holds approximately 15% of the volume of the entire gastro-intestinal tract. The stomach is a thin walled and has little power of contraction. Stomach has three parts-fundus, antrum and pylorus. The well-developed cardiac sphincter which is lined with stratified squamous epithelium, prevents vomiting. The normal pH within rabbit stomach is highly acidic and varies from 1.0 to 2.0. The stomach pH of rabbits until the time of weaning is around 5.0–6.5. During this period, bacteria is kept in check by the production of milk oil containing octanoic and decanoic fatty acids produced by the enzymatic reaction on the doe's milk. The food remains in the acidic stomach for three to six hours, undergoing little chemical change.

The gastric contents are slowly released into the small intestine for further digestion. The small intestine is approximately 12% of the gastrointestinal volume in the rabbit. The food materials pass from the stomach through a muscular band of tissue, called the pylorus, which controls the entry of the food into small intestine. The first part of intestine, duodenum, is loop like structure within which pancreas is present. The right lobe of the pancreas is situated in the mesoduodenum of the duodenal loop whereas the left lobe lies between the stomach and transverse colon. There is a single pancreatic duct which drains both pancreatic lobes. Pancreas supplies certain fluids to duodenum which has two functions: (*a*) to provide enzymes for digestion, and (*b*) to control sugar levels in the tissues and bloodstream. The liver is located under the diaphragm and performs many functions regarding metabolism, digestion and detoxification. The bile duct, coming from the gall bladder in the liver, is entered in the duodenum which carries bile salt containing biliverdin rather than bilirubin. Bile and pancreatic enzymes, which are released into the small intestine, break down protein and fat into very small minute particles suitable for absorption. The lower parts of small intestine are known as jejunum and ileum. The small intestine of rabbit extends upto the enlarged sac or sacculus rotundus. The normal pH with in small intestine is around 7.2. The distal end of the ileum has a spherical thick-walled enlargement known as the sacculus rotundus and this structure is often called the "cecal tonsil" because of its lymphoid tissue and macrophage composition. This organ is unique to rabbits. An ileocolic valve controls movement of ingesta from the ileum into the large intestine and prevents reverse movement of ingesta back up into the ileum. The small intestine is connected with large intestine through ileocaecal valve.

The large intestine consists of the caecum, the colon and the rectum. The cecum of the rabbit is large and may contain 40% of intestinal content. It has 10 times the capacity of the stomach. The caecum meets the colon at the sacculus rotundus. At the end of caecum, appendix is attached. This appendix contains lymphoid tissue and secretes bicarbonate that buffers the cecal acids, and water to form the ccecal paste. The fermentation process in the cecum results in volatile fatty acids that are absorbed across the cecal epithelium. Young rabbits get gut flora by consumption of the doe's cecotrophs at 2 weeks of age. The bovine's main volatile fatty acid (VFA) produced by rumen fermentation is propionic acid while the rabbit's main VFA is acetic acid with cecal fermentation. In addition other VFA like acetate, butyrate, and propionate are also produced. The primary microflora of the rabbit is *Bacteroides* species while *Lactobacillus* species is the primary microflora of the bovid. In addition to *Bacteroides* species, there may also be ciliated protozoa, yeasts, and small numbers of *E. coli* and clostridia species in the cecal flora of rabbit. Cecal contents have an alkaline pH in the morning and an acid pH in the mid afternoon due to flacutation of microorganism population and this is termed as "transfaunation".

The colon has proximal and distal parts where pH is around 6.5. The final part of large intestine is rectum which opens through anus. The ascending colon is divided into 4 sections. The ampulla coli opens into the first section, approximately 10 cm long and having 3 longitudinal flat bands of muscular tissue (taeniae). The section has small protrusions (about 0.5 mm in diametre) that are named "warzen" or warts. This are unique feature to lagomorphs. These warts greatly increase the surface are of the colon for absorption and also aid in mechanical separation of ingesta. The second section of colon has a single taenia and fewer, smaller haustra. The contractions of haustra mechanically separates the ingesta into indigestible particles and liquid contents. As the large pellets pass down the middle of the lumen, water is re-absorbed and they are excreted as hard dry pellets. The third section is the fusus coli. It is a muscular area about 4 cm long, highly innervated, and vascular. Its mucosal surface has prominent longitudinal folds and goblet cells. It opens into the fourth section of colony. The distal colon ends at the rectum. Its mucosa has short crypts with abundant goblet cells. It is thin-walled and usually contains hard fecal pellets.

Cecotrophy

Caecotrophy, commonly known as coprophagy, is a specialized digestive strategy common in the Mammalian order Lagomorpha (rabbits, hares and pikas) and some species of rodents and it plays a central role in digestive physiology. It is the ingestion of specially produced soft faecal pellets, which are usually produced at night in the domestic rabbit. Rabbits begin consuming caecotrophs between 2 and 3 weeks of age as they begin to eat solid food. Caecotrophy in small herbivorous mammals is considered an adaptation to the metabolic disadvantage of small body size, especially when feeding on low quality diets. When a small mammal ingests food rich in plant fibre, it is retained in the digestive tract for a long period of time. The bulk in the stomach effects intestinal passage of digesta. No fresh food can be consumed because of the lack of storage volume in the digestive tract, and as a result the animal starves to death, although its digestive tract is full (Langer, 2002). Compared to the ruminant, the rabbit is much less efficient at digesting fibre. In ruminant, microbial activity in the ruminant affects the food before it reaches the main regions of digestion and absorption, but in the rabbit, microbial activity takes place on the undigested residues of food. The rapid metabolic rate of small herbivores requires that energy which has to be extracted rapidly and food should not be retained in the gastrointestinal tract for long periods. Through caecotrophy small mammals are able to ingest a "home-made" food rich in microbial products, such as protein and vitamins. It is similar to the chewing of cud in cattle, and is sometimes referred to as "pseudo-rumination". However, the two processes are very different.

Cecotrophs are formed in the proximal colon and cecum. Fiber material greater than 0.5 mm does not enter the cecum but transits to be formed and passed as hard fecal pellets. The smaller particles and fluid remain in the cecum or are returned to the cecum via antiperistalsis to form high nutrient particles that become coated with mucus as they pass through the colon and they are usually passed 8 hours or so after feeding, which coincides usually to night time. Cecotroph digestion and some fermentation take place during 6–8 hours when they remain in the gastric fundus. Cecotrophs contain microorganisms and products of microbial fermentation including amino acids, volatile fatty acids, and vitamins. A gelatinous mucous coating protects them from some of the stomach acid. As the cecotrophs passed through the colon, lysozyme was incorporated. The lysozyme has bacteriolytic activity that degrades microbial proteins for absorption in the small intestine. Bacteria within the cecotroph produce amylase that converts glucose to carbon dioxide and lactic acid. These products along with amino acids and vitamins are absorbed primarily in the small intestine. This mechanism requires high fiber diets to function properly. Low fiber diets increase cecal retention time and promote hypomotility of the entire gut, which further reduces the production of cecotrophs. Fiber in the diet should be indigestible and it should constitute at least 15% of total food. A low protein diet increases a rabbit's cecotroph ingestion. A high protein diet but low in fiber reduces consumption. In crude fiber terms, diets that are less than 150 g/kg of feed will almost always result in digestive upset while diets with greater than 200 g/kg crude fiber result in increased incidence of cecal impaction and mucoid enteritis. A diet devoid of fiber has a coefficient of apparent digestibility of organic matter of 0.90. This declines in a linear fashion to 0.40 when the diet contains 350 g crude fiber per kilogram of feed. Increased crude fiber of the diet increases the crude fiber of the cecal contents. This decreases the protein content. Compounded and pelleted diets require addition of hay in order to supply a complete diet. In general, the recommendation is that hay should be supplied on a free-choice basis as a rule of good husbandry of the pet rabbit should be emphasized.

MALE REPRODUCTIVE SYSTEM

The male reproductive system consists of testis, epididymis, vas deferens, urethra and penis, and accessory glands *i.e.*, prostate, Cowper's and perineal glands (Fig. 9). The reproductive organs of the male rabbit are very similar to that of other mammals except for the additional ability to retract the testis in the abdomen. Male rabbits possess a life-long ability to move freely their testis between the scrotum and the abdomen, *via* the inguinal ring. The response occurs during handling and during a fight between two males. Male rabbits will inevitably fight and attempt to "neuter" the other male by biting in the scrotal region.

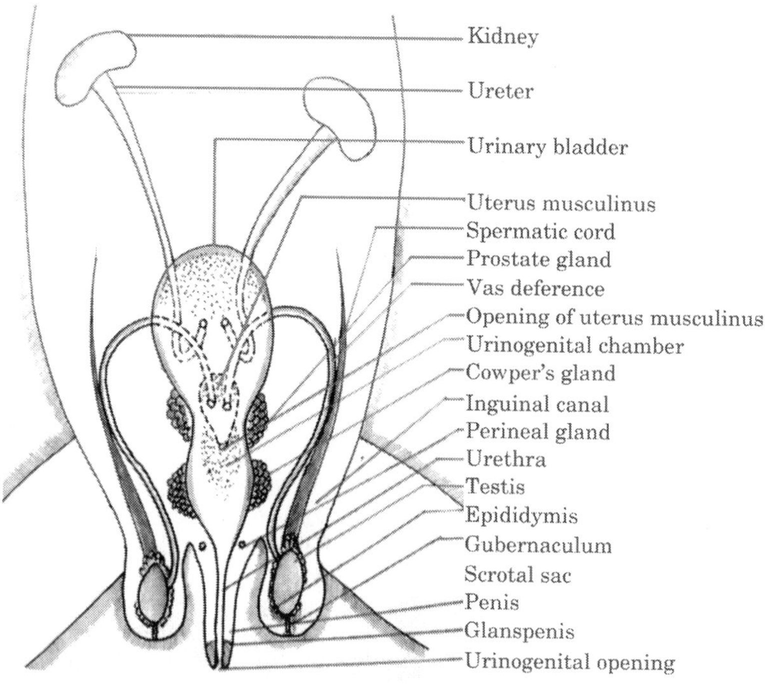

Fig. 9: Male rabbit reproductive system

Testes

The testes descent into the scrotum at the age of 2.5–3 months. Rabbit has two testicles situated on the either side of urogenital opening. It is held in its positions within an extension of abdomen, called scrotum. Each one is supported from posterior scrotal wall by means of a small cord of fibro muscular mesodermal tissue, the gubernaculum. Testes of rabbit (Fig. 10) are about 15–20 mm in length, 10 mm in thick, oval in shape and pinkish in colour. Under a tough membranous envelope, the tunica albuginea, each testis contains slender and highly coiled seminiferous tubules (Fig. 11). These tubules are embedded in a loose interstitial

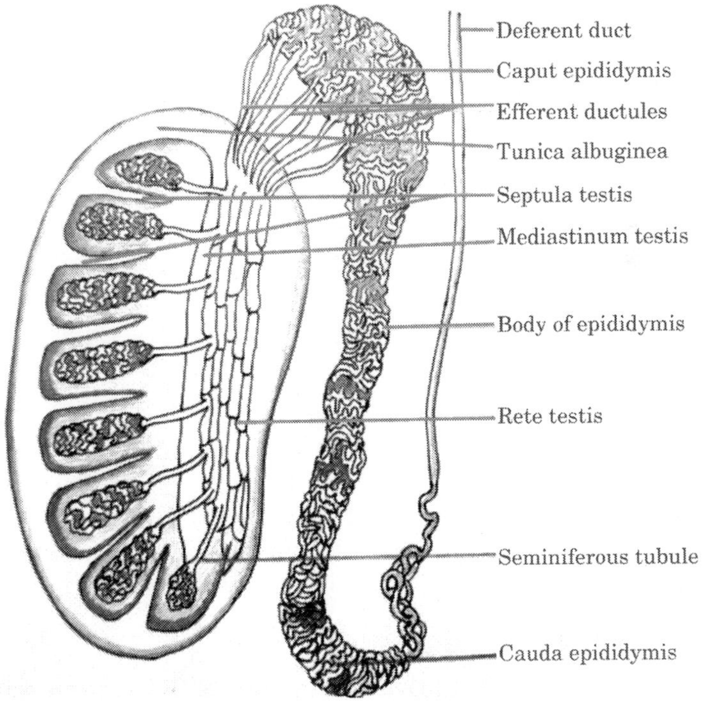

Fig. 10: Rabbit testis and epididymis

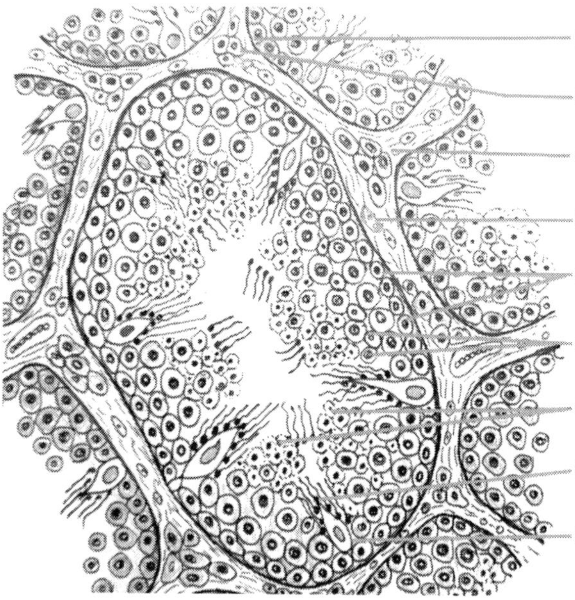

Fig. 11: Enlarged section of seminiferous tubule

connective tissue containing blood vessels, nerve fibres and glandular cells secreting male sex hormones. The seminiferous tubules produce spermatozoa. The sperm then enters into *rete testis* located in the mediastinum testis, to the efferent ducts, and ultimately, into epididymis where the sperm cells mature.

The epididymis, a compact, narrow, somewhat flattened and whitish tightly coiled mass representing the Wolffian duct, is closely abutted against the dorsal aspect of each testis. This mass can be divided into three parts: (*a*) anterior enlarged part or caput epididymis, (*b*) middle narrow or corpus epididymis, and (*c*) relatively smaller posterior enlarged part or cauda epididymis. The total length of epididymis is about three metres. The Wolffian duct continues forward as an uncoiled and to some extent thicker male genital duct which is called vas deferens. It passes through a spermatic cord comprises of fibro muscular tissue, artery, vein, nerve and lymphatics and enters into the abdominal cavity. Each vas deferens runs along dorsal aspect of urinary bladder, turns back after forming loop around the ureter of its side, and opens into uterus masculinus near the base of urethra.

Uterus masculinus, a glandular pouch like diverticulum, secretes thick and sticky alkaline fluid which is a part of semen. The anterior part of urethra is narrow and opens at the tip of penis by a urinogenital aperture. Penis is a finger like cylindrical growth of abdomen covered by skin projecting in between the scrotal sacs.

The tip of penis, glans penis, is swollen and smooth. The skin is folded at the base of glans which is called prepuce. It is the copulatory organ of male. During sexual excitation the penis is erected, becomes rigid, longer and thicker. At the time of copulation the erect penis enters into vagina of the mating female to discharge semen.

Accessory Glands

1. ***Prostate gland:*** It consists of a number of lobules which open near uterus musculinus and vasa differentia through small ductule. This gland secretes white milky alkaline fluid containing enzymes, fructose and certain proteins. The mixing of fluids from prostate and musculinus results coagulation. It gives semen thick and viscous appearance.
2. ***Cowper's gland or Bulbourethral gland:*** They are two in numbers and are located on either side of urethra before entering into the penis. The glands secrete a transparent and alkaline fluid at the time of sexual excitation to lubricate the passage of semen, to neutralize the acid remaining in the urethra due to previous discharge of urine and to lubricate the vagina

of female counterpart facilitating copulation as well as sperm protection from abrasion against vaginal wall.

3. ***Perineal glands*:** These glands are present within the perineal pouches which are located in either side of the base of penis. The secretion of these glands attracts females.
4. ***Rectal glands*:** These glands are found near the rectum whose smell attracts females.

FEMALE REPRODUCTIVE SYSTEM

The female reproductive system comprises with ovary, fimbrae, fallopian tube, uterus, vagina and accessory glands *i.e.*, Bartholin's glands, perineal and rectal glands (Fig. 12). The reproductive organ of the female rabbit is duplex: the uterus is formed by two independent horns, split over their whole length. Each horn possesses its own cervix. The split two-horned system is only observed in monotreme egg lying mammals and in lagomorphs (pika, hare and rabbit).

Ovary

Two ovaries are smaller in size (maximal length of 1–1.5 cm), whitish in colour and oval in shape and they are located at the end of the uterus, right under the kidneys and suspends from the dorsal wall of body through mesovarium. Very close to each ovary and also supported by mesovarium, there is a funnel-like infundibular structure with a numbers of thickly ciliated fingers-like border called fimbrae. The funnel leads to a very narrow thin walled and less convoluted tube which is known as fallopian tube. Internally, this tube is lined with ciliated epithelium. The tubes from both the ovaries obliquely extend backwards and medianwards, and enlarge before reaching the median line to form a thick walled structure is called the uterus. The uteri of both sides open in a wide, pouch-like structure to

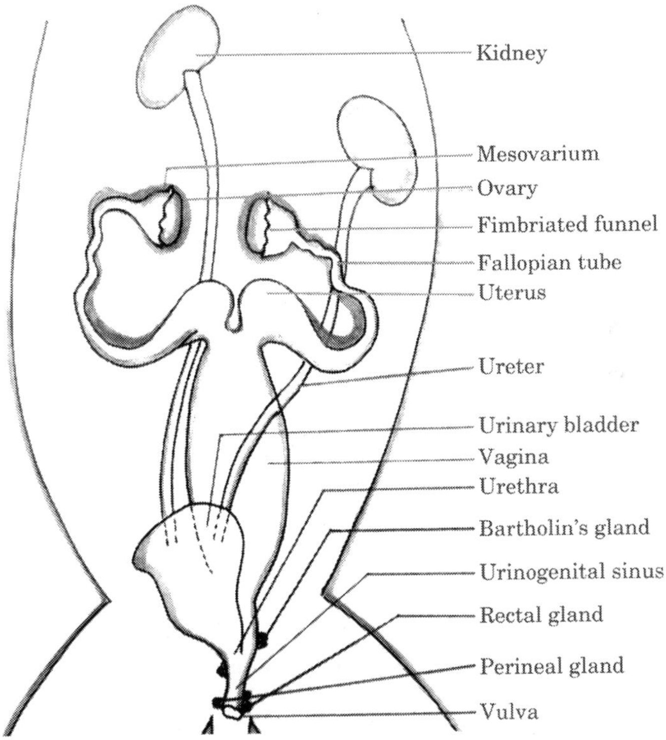

Fig. 12: Female reproductive system of rabbit

form vagina which is located dorsally to urinary bladder and ventrally to rectum. At the posterior side, the vagina becomes narrower and receives the short, narrow and neck-like continuation of urinary bladder the urethra is called urogenital sinus or vestibule. The vestibule opens at the hind end of the body by a ventral, slit-like aperture which is known as vulva.

Sex differentiation in rabbit occurs during the embryonic phase, on the 16th day post fertilization. The ovaries develop from the cluster of cells lying near the original testes. The development of the ovaries is associated with the degeneration of the testes. The development of the ovules (female reproductive cell) starts around the 21st day and continues till birth, around the 30th day. The first ova and follicles start to develop only 13 days after birth. The transverse section of rabbit ovary is shown in Fig. 13. Ovaries are made up of a dense and fibrous connective tissue, the stroma. Each has an envelope comprised of an outer peritoneum, middle germinal epithelium, and an inner connective tissue membrane, the tunica albuginea. A number of small ovarian or graafian follicles are found embedded in the cortical stroma of an ovary which is formed by proliferation of germ cells of germinal epithelium. Initially, this follicle is known as primordial follicle. It remains in dormant condition during sexual maturation. Few of them degenerates and few become active during breeding season. Over the period of time the primordial

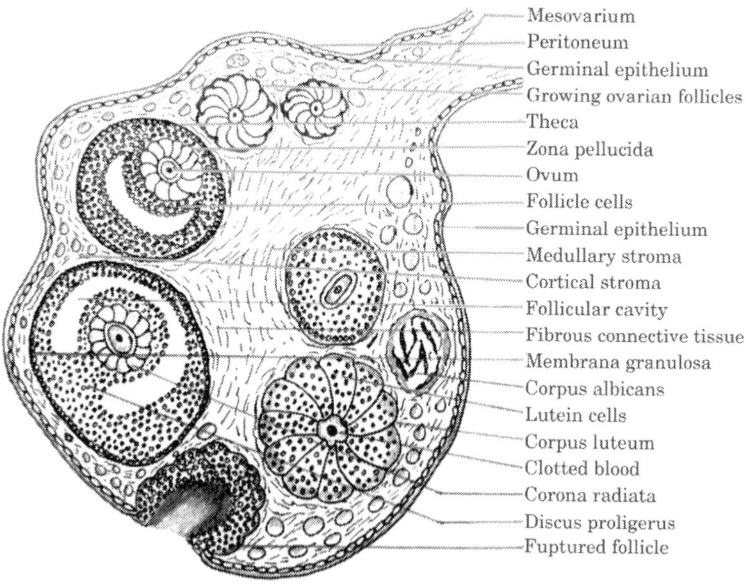

Fig. 13: Transverse section of ovary in rabbit

follicles turn into primary follicles, secondary follicles and ultimately, graafian follicles. Developing graafian follicle gradually presses outwards forming a bulge at ovarian surface and ruptures. This phenomenon is known as ovulation. Then, the secondary oocyte present in the follicle directly falls into the fimbriated funnel of oviduct. The ovulation in rabbit occurs only when copulation starts. It is a reflex action related to coitus. That is why, the female rabbit is known as 'reflex ovulator' or 'induce ovulator'. The rupture follicle is changed into yellowish endocrine gland or corpus luteum which secretes progesterone. Later, this progesterone maintains pregnancy in animals.

REPRODUCTIVE ORGANS OF FEMALE RABBIT

Accessory Glands

1. ***Bartholin's gland*:** These glands are located between vestibule and rectum. They secret a slimy, alkaline fluid into the vestibule to lubricate for facilitating copulation.
2. ***Perineal glands*:** Just like male rabbit, these glands are present within the perineal pouches of either side of vulva. The glands secret some chemicals which attract male rabbit.
3. ***Rectal glands*:** The location of these glands are same as male rabbit and the smell of this gland's secretion attracts the counterpart.

RESPIRATORY SYSTEM

The rabbit, a terrestrial animal, respires through lungs which are located in their thoracic cavity. The process of respiration especially gaseous exchange is highly complicated phenomenon. Like other mammals the respiratory system of rabbit consists of following parts:

1. ***External nostrils or external nares:*** Rabbit has two oval shaped external nostrils located at the front end of its snout. They open into nasal passages.
2. ***Nasal chambers:*** These are paired long passages above the palate separated from each other by a thin, cartilaginous median vertical partition called nasal septum. This chamber can be divided into three parts- (*a*) anterior vestibular part, (*b*) middle respiratory part, and (*c*) posterior olfactory part.

   Vestibular parts are within the external nostrils and are lined by stratified squamous cell epithelium. This region contains hair which acts as sieve and has sebaceous glands. Respiratory parts form the major portion of nasal chambers. They contain three thin, twisted bony plates *i.e.*, superior, middle and inferior 'turbinals' or 'conchae'. This part has highly vascular, lined by ciliated columnar epithelium, and has numerous mucous cells. The secretions of these cells keep the conchae moist. Thus, this part acts as natural conditioners. Olfactory parts lie in the membrane lining the superior nasal conchae and adjacent septum. There is a yellowish-brown, nonciliated and columnar neuroepithelium comprised of bipolar olfactory neurons, supporting subtentacular cells and basal cells. This epithelium, serves for perception of smell, is also called olfactory epithelium or Schneiderian membrane.
3. ***Internal nostrils:*** Nasal passages open into the nasopharynx through a pair of internal nostrils located above the soft palate.
4. ***Pharynx:*** This is a common passage for passing air and food which cross each other in it. Nasopharynx is lined by pseudostratified ciliated epithelium while laryngopharynx is by non-keratinised stratified squamous epithelium. At the posterior end of laryngopharynx, there is a slit like pore which is known

as glottis. It has a lid called epiglottis made up of elastic cartilage. It closes the glottis during swallowing of food or water. Otherwise, glottis remains open into larynx.

5. ***Larynx*:** This part is a short, tube like structure and is supported by four cartilages *viz.*, a thyroid, a cricoid and a pair of arytenoids. Thyroid cartilage, the largest one, forms the ventral and lateral walls of the larynx and incomplete dorsally. Epiglottis arises from the thyroid cartilage. The cricoid cartilage is situated behind the thyroid while a pair of arytenoid cartilages is on the dorsal side of larynx. The tips of arytenoids in their turn bear a pair of cartilages which is known as snatorini. There are two fibro elastic strands extending between thyroid and arytenoids cartilages are called vocal cords. The space between two vocal cords is called rima glottides. Sound is produced as a result of their vibrations. Therefore, larynx is known as voice producing organ. This larynx opens into the trachea.

6. ***Trachea or wind pipe*:** This is a thin walled, wide tube passes through the ventral side of the oesophagus of neck. It is supported by C-shaped cartilaginous rings, which incomplete dorsally. Each ring is enclosed in perichondrium. All rings are embedded in a fibrous connective tissue membrane, rich in collagen, elastin and smooth muscle fibres. This membrane encloses each ring, connects it with adjacent rings and tracheal wall and fills up the gap between its free dorsal ends. Inner layer of trachea consists of pseudostratified ciliated columnar epithelium cells. Thus, it becomes non collapsible and to some extent stretchable.

7. ***Bronchi*:** After entering into thorax, trachea bifurcates into two primary bronchi. Each primary bronchus enters into the corresponding lung. Bronchi are lined by pseudo stratified ciliated columnar epithelium with goblet cells. Mucous secreted by these cells traps the dust particles

and bacteria. Within the lung, each primary bronchus is divided into two secondary bronchi and further tertiary bronchi.

8. ***Bronchioles*:** Tertiary bronchi are again divided into many branches called bronchiole, which are in the order primary, secondary, tertiary, terminal and respiratory bronchioles (Fig. 14). Larger bronchioles are lined by simple ciliated columnar epithelium and smaller by simple non-ciliated cuboidal epithelium. Each respiratory bronchioles are again divided into respiratory terminates in a cluster of alveolar ducts. They are ended in thin walled and highly vascular alveolar sacs. Each alveolar sac is formed of many small round or oval chambers called alveoli. The gaseous exchange takes place in these places.

9. ***Lungs*:** Rabbit has a pair of pinkish and spongy lungs in the thoracic cavity, one on either side of heart (Fig. 15). Each lung is enclosed in a double layered pleuroperitoneum in which outer layer is known as

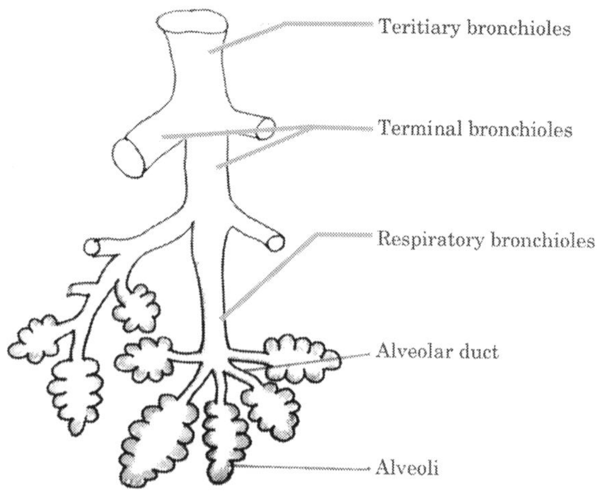

Fig. 14: Respiratory passage

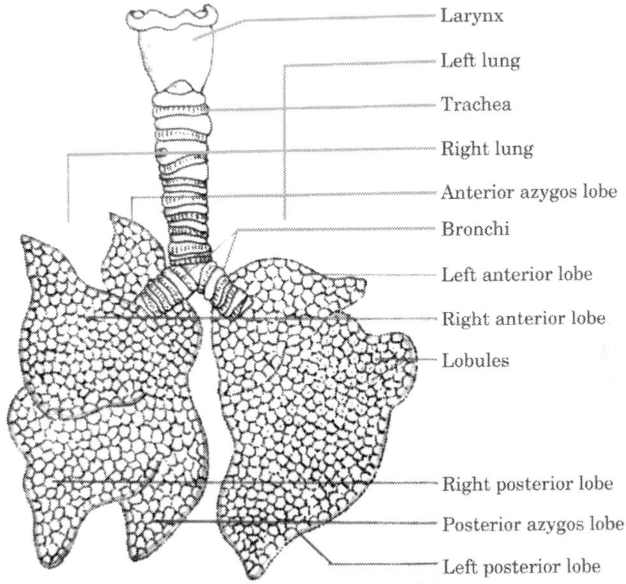

Fig. 15: Respiratory system of rabbit

parietal layer and the inner layer is called visceral layer. These two layers are separated by pleural cavity which contain pleural fluid. This fluid protects the lungs from shock and friction. Right lung of rabbit has four lobes *viz*., anterior azygous, right anterior lobe, right posterior lobe and posterior azygous lobe. Left lung has two *i.e.*, left anterior and left posterior lobes.

Chapter 2

Breeds of Rabbits

There are two types of rabbit breed– (*i*) Domestic rabbit breeds and (*ii*) Wild rabbit breeds. Most of today's domestic breeds are originated from the European rabbit through selective breeding (cross breeding) or natural selection. Development of breed is literally a never ending and continuous process. The breeds are recognized either by American Rabbit Breeders' Association (ARBA) or the British Rabbit Council (BRC). Although there are big numbers (more than 130, Table 4) of breeds but according to ARBA there are currently 47 different recognized rabbit breeds.

CLASSIFICATION OF RABBIT BREEDS

The rabbits are used for wool and meat production, as model laboratory animals for biomedical research work and for exhibition purpose. Depending upon the utility, rabbit breeds can be classified into four different categories: (*a*) Wool breeds *e.g.*, Angora, New Zealand White; (*b*) Meat breeds *e.g.*, Chinchilla, Satin; (*c*) Fur breeds *e.g.*, Himalayan, Lilac; and (*d*) Show breeds *e.g.*, Lop, Dutch etc. Today, more than 47 recognised breeds are available all over the world which is different from each other in terms of phonotypic characteristics as well as their end use. Some of the internationally recognised breeds (Fig. 16) are described in this chapter.

Table 4: Breeds of rabbit

Origin	*Breed*	*Body weight (kg)*	*Colour(s)*
Algeria	1. Kabyle	2.3–2.7	Many
Asia	2. Himalayan	2.7–3.6	White with black, Chocolate, Lilac or Blue points
Australia	3. Enderby Island	1.4–1.8	Champagne, Crème
Belgium	4. Belgian Hare	3.6–4.5	Black, Black & Tan, Red
	5. Beveran	2.3–3.6	Black, Blue, White
	6. Blanc de Termonde	4.5–5.4	White, Red Eyes
	7. Blue of Ham	4.5–5.9	
	8. Blue of Sint-Niklaas	2.3–5.4	Blue
	9. Flemish Giant	6.4	Steel, Grey, Sandy
	10. Grey Pearl of Halle	2.3–2.7	Light Grey
	11. Petite Brabancon*	—	—
	12. Pointed Beveran	3.2	Blue points, Brown or Lilac
Brazil	13. Brazilian	3.2–5.0	
Canada	14. Satin Angora	2.7–4.1	
Central America	15. Cuban Brown		Chocolate
Czech Republic	16. Czech Albin		White
	17. Czech Red rabbit	1.8–2.3	Chestnut
	18. Czech Spot	2.7–3.6	Black, Blue, Agouti, Tri-color

Table 4: (*Contd...*)

Table 4: (*Contd...*)

Origin	*Breed*	*Body weight (kg)*	*Colour(s)*
	19. New Zealand Red	4.1–5.4	Chestnut
Egypt	20. Baladi	2.7	Black, Red, White
	21. Bauscat	3.6	Albino
	22. Gabali	2.7–3.2	Agouti
	23. Giza White	2.3–3.2	White (albino)
England	24. Britannia Petite	0.68–1.1	Black, Black Otter, Chestnut, Sable Marten, Red-Eyed White
	25. British Giant	5.4–5.9	Black, Blue, White, Opal, Steel Grey, Brown Grey
	26. Cashmere Lop	1.8–2.3	Many
	27. Dutch	1.8–2.3	Black, Blue, Chocolate, Grey, white nose and white paws
	28. Dutch (Tri-coloured)	1.8–2.3	Tri-colour with a white band across shoulders, white nose and paws
	29. English Lop	4.5–5.0	Many
	30. English Spot	2.3–3.6	White with colored butterfly pattern
	31. Golden Glavcot	2.3–2.7	
	32. Lilac	2.3–3.2	Lavender

Table 4: (*Contd...*)

Table 4: (*Contd...*)

Origin	*Breed*	*Body weight (kg)*	*Colour(s)*
	33. Mini Lion Lop	0.91–1.4	Many
	34. Old English Red*	—	—
	35. Perlfee	2.3–3.6	Blue Agouti
	36. Polish	0.45–0.91	Many
	37. Rex (Astrex)	2.7–4.5	Many
	38. Rex (Mini)	1.4–1.8	Many
	39. Rex (Opossum)	2.7–3.6	Many
	40. Sitka*	—	—
	41. Sussex	3.2–3.6	Gold, Cream
	42. Tan	1.8–2.7	Black, Blue, Chocolate, Lilac
England/ Netherlands	43. Beige	2.9	Dark Chamois or Light Sand
Europe	44. Checkered Giant	5.0–5.4	White with black markings
	45. Continental Giant	5.4–7.3	Many
	46. Giant Papillon	5.0–5.4	White with black markings
	47. Patagonian*	—	—
France	48. Argente Blue	2.7	Blue
	49. Argente Brun	2.7	Brown
	50. Argente Clair	2.7	Blue Silver

Table 4: (*Contd...*)

Table 4: (*Contd...*)

Origin	*Breed*	*Body weight (kg)*	*Colour(s)*
	51. Argente Crème	2.7	Cream
	52. Argente de Champagne	3.6–4.5	Silver
	53. Argente Noir	2.7	Dark Silver
	54. Argente St Hubert	2.7	Silvered Black Agouti
	55. Blanc de Bouscat	5.0–6.8	White
	56. Blanc de Hotot	3.6–5.0	White, dark rings around eyes
	57. Blanc de L'Oural*	3.6–4.5	Albino
	58. Blanc de Popielno	4.5–5.4	White
	59. Bourbonnais Grey	3.2–5.0	Blue
	60. Brown Chestnut of Lorraine	2.0	Chestnut
	61. Chaudry	3.6–4.1	Red-Eyed White
	62. Chinchilla (Giganta)	4.5–5.4	Chinchilla
	63. Chinchilla (Standard)	2.3–3.2	Chinchilla
	64. English Angora	2.3–3.6	Many
	65. Fauve de Bourgogne	3.2–5.0	Orange/Red
	66. Fee de Marbourg(Marburger)	1.8–3.2	Lilac
	67. French Angora	3.2–4.5	Many
	68. French Lop	4.5	Many
	69. Harlequin	2.7–4.1	2 colours in fur *i.e.*, part black part orange

Table 4: (*Contd...*)

Table 4: (*Contd...*)

Origin	*Breed*	*Body weight (kg)*	*Colour(s)*
	70. Rex (Standard)	2.7–4.5	Many
	71. Rouennais*	3.6–4.5	—
Germany	72. Alaska	3.2–4.1	Black
	73. Dwarf Hotot	0.91–1.4	White with black around the eyes
	74. Dwarf Lop (Mini Lop in USA)	1.8–2.3	Many
	75. German Angora	1.8–5.4	White, Albino
	76. German Lop	2.7–3.6	Many
	77. Meissner Lop	4.5–5.4	Many
	78. Rhinelander	4.1–4.5	White with coloured butterfly patterns
	79. Sachsengold		Chestnut
	80. Teddywidder	1.5	Many
	81. Thuringer	4.1	
Hungary	82. Hungarian Giant	5.4–6.8	Many
	83. Pannon White	2.3–4.5	White
Isle of Man	84. Swan*	7.3–9.1	Brown, Grey
Italy	85. Carmagnola Grey	4.5–5.4	Chinchilla
Japan	86. Pani		
Mexico	87. Criollo		

Table 4: (*Contd...*)

Table 4: (*Contd...*)

Origin	*Breed*	*Body weight (kg)*	*Colour(s)*
Morocco	88. Tadla	1.8–2.3	Agouti
	89. Zemmouri		
Netherlands	90. Deilenaar	2.3–3.6	Red Agouti
	91. Havana	2.0–2.9	Chocolate, Blue, Black, Lilac
	92. Hulstlander	1.8–2.7	White with blue eyes
	93. Miniature Lop (Holland Lopin USA)	0.91–1.4	Many
	94. Netherland Dwarf	0.50–1.6	Many
	95. Sallander	2.3–4.1	
Scandinavia	96. Orestad	2.3–3.2	Albino
Slovakia	97. Liptov Baldspotted Rabbit	3.6–4.1	Agouti, Black
Spain	98. Caldes		Red-Eyed White
	99. Spanish Giant	5.7–6.8	
Sweden	100. Elfin	1.8–2.3	Many
	101. Gotland	1.4–1.8	
Swiss Alps	102. Thrianta	2.7	Chestnut
Turkey	103. Güzelçaml Rabbit	2.3–3.6	White with chocolate markings
United States			
	104. Altex	5.9	Albino

Table 4: (*Contd...*)

Table 4: (*Contd...*)

Origin	*Breed*	*Body weight (kg)*	*Colour(s)*
	105. American Blue	4.1–5.4	Blue
	106. American Fuzzy Lop	1.4–1.8	
	107. American Sable	4.1–4.5	Sable
	108. American White	4.1–5.4	White
	109. Californian	4.1–4.5	White with Chocolate, Lilac or Blue points, Red eyes
	110. Chinchilla (American)	4.1–5.4	Chinchilla
	111. Chinchilla (Giant)	4.5–7.3	Chinchilla
	112. Cinnamon	4.5–5.0	Cinnamon
	113. Florida White	1.8–2.7	White
	114. Giant Angora	4.5	
	115. Jersey Wooly	1.1–1.6	Many
	116. Lionhead	1.1–1.6	
	117. New Zealand	4.1–5.4	White with red eyes, Red, Black, Broken
	118. Palomino	4.5–5.0	Lynx, Golden
	119. Plush Lop (Mini)	1.4–1.8	
	120. Plush Lop (Standard)	2.3–3.6	
	121. San Juan	1.4–2.3	Agouti

Table 4: (*Contd...*)

Table 4: *(Contd...)*

Origin	*Breed*	*Body weight (kg)*	*Colour(s)*
	122. Silver Fox	5.0–5.4	
Other Countries	123. Blue Beveran*	9	Blue Dutch
	124. Fox Silver	2.3–2.7	Black, Lilac, Chocolate, Blue
	125. Satin	2.7–3.6	All self colours, Very shiny fur
	126. Satin (Mini)	1.8–2.3	
	127. Siamese Sable	2.3–3.2	
	128. Siberian	2.3–3.2	
	129. Silver	2.3–3.2	
	130. Silver Marten	4.3	
	131. Smoke Pearl	2.3–3.2	
	132. Squirrel	2.3–3.2	
	133. Swiss Fox	2.3–3.6	
	134. Vienna	3.2–4.1	Blue, Black, Agouti, White
	135. Wheaten	2.5–3.2	Fawn or Pale yellow
	136. Wheaten Lynx	2.5–3.2	Orange shot silver
	137. Beaver*	—	Silky & Luxurious coat
	138. Blanc de Chauny*	—	Snow white
	139. Golden Fawn*	—	—

* These breeds are now extinct but many contribute in the inheritance of present rabbit breeds

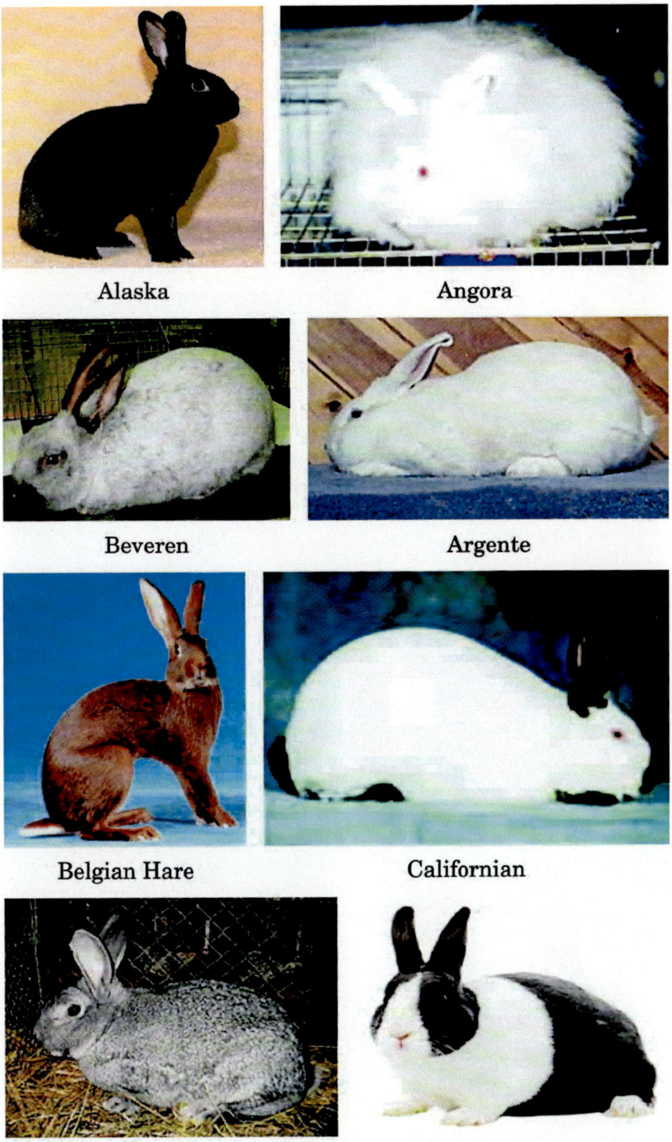

Fig. 16: Rabbit breeds (*Contd...*)

Flemish Giant

Harlequin

Havana

Himalayan

Lilac

Lop

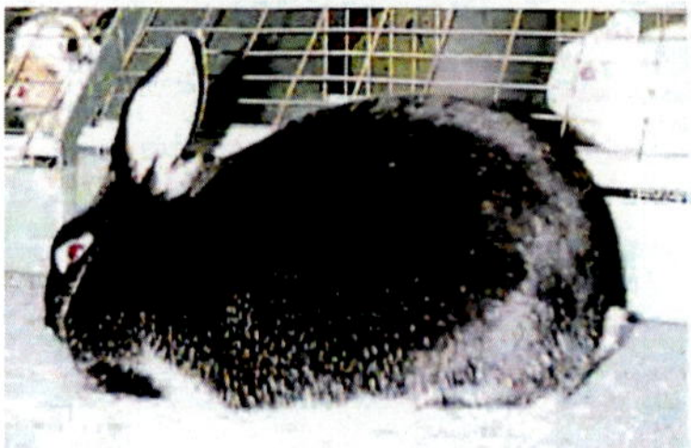

Marten

Netherland Dwarf

Fig. 16: Rabbit breeds (*Contd...*)

New Zealand White

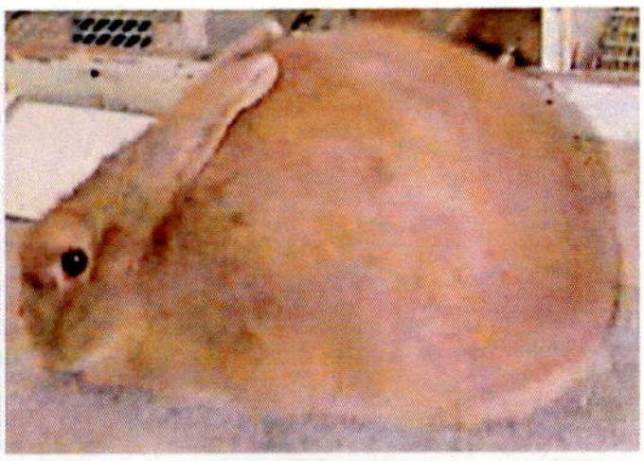

Palomino

Polish

Rex

Rhinelander

Sable

Satin

Siberian

Fig. 16: Rabbit breeds (*Contd...*)

Smoke Pearl

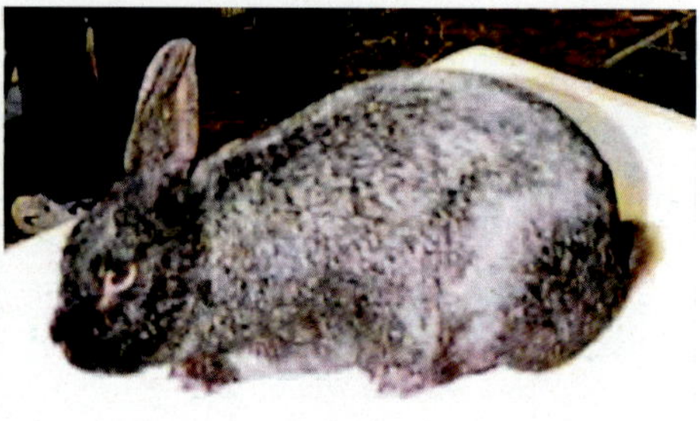

Fox

Thrianta

Vienna

Fig. 16: Rabbit breeds

ALASKA

The Alaska breed was developed in Germany in the later years of the 1920's from Himalayan and Argente breed. It was raised for its gorgeous fur but now, is used as pet and show purposes. The body size is medium and varies from 3.0 to 4.0 kg. This rabbit is stumpy, thickset and has rounded body. They possess broad head, almost no neck, very close to the body. Doe has comparatively finer neck with well developed dewlap than buck. The shoulder, chest, rump and legs are very strong. The coat is silky, thick, lustrous and jet black in colour. The eyes are bright, well opened, brownish black and nails are black.

ANGORA

This is one of the oldest breed of rabbits and is thought to be originated in Ankara, the capital of Turkey. This breed is very docile, calm and popular for its soft, silky and long hair which resembles a fur ball. Basically, this breed is white in colour but breeders have developed many coloured Angora breeds. Long and white staple length has more demand in the garment industry due to easy and uniform dyeing of wool. The Angora rabbit can produce average 3–5 young in one litter. Litters of up to 6 or even more have been recorded. There are many individual breeds of Angora breeds *viz.*, German, English, French, Russian, Giant, Satin etc.

German

This ruby-eyed white breed is heaviest among other Angoras. The adult body weight ranges from 3 to 5 kg. Due to the well developed foot gland, this breed is fittest in wire cage rearing under intensive system. They produce pure white, finest, softest, lusturous long coats with 0–3% guard hair which is referred to as 'wool'. They will need regular grooming to keep their fur in good condition and hay as a fibre source of regular part of their diet to help prevent hair balls in their digestive tract. The fibre diameter is about 12–16 micron. This breed has the capability to produce annually upto 1200 gm of wool per animal under very good climatic condition with optimum management. In tropical climate, annual clip per animal is around 550 to 700 gm. However, their wool is very famous for making warmth garments and is sold in high prices in the market.

English

This is a smallest Angora breed having adult body weight of 2.0–3.5 kg. The eye is large and round. They have very long heavy fur that covers it's body except face and resembling the feature of a small dog. The floppy tassels of silky and fine wool lining are also present in the ear. The annual wool yield from one animal is around 500 gm with fibre diameter of 12–16 micron and guard hair of 4–6%.

French

The adult body weight varies from 3.0 to 4.5 kg. This breed has clean head, ear, face and front feet and rest of the body has a preponderance of guard hair with undercoat wool. Thus, they are comparatively easy to maintain. The annual wool yield varies from 500–800 gm per animal.

Russian

This is a comparatively larger breed among Angoras. The adult body weight varies from 3.5 to 4.5 kg. They produce coarse wool with 16-18 micron fibre diameter and 14–15% guard hair. The average annual wool yield is very poor *i.e.*, 300–400 gm per animal. Russian Angora breed were imported from USSR to India during 1978–79 at the Division of Fur Animal Breeding, North Temperate Research Station, Garsa, under Central Sheep and Wool Research Institute (CSWRI), Avikanagar.

Satin

The Satin Angora is a cross between Satin and French Angora weighing 3.0 to 4.5 kg. They have clean head, face, ears and front feet like French Angora. The wool is extremely soft in texture but the annual yield is very less due to less density of fibre.

Giant

The Giant Angora has been developed by crossing between German Angora and other large size short haired rabbits for higher wool production. German Angora and Giant are the two different breeds. The adult body weight of this breed is 4.5 kg or more. The coat colour is ruby eyed white. In addition to the undercoat and guard hair, they have 'Awn Fluff' that does not exist in other three breeds of Angora. The Giant Angora does not molt like other rabbits, therefore, require clipping twice in a year. Their growth rate is very first comparative to other 'giant' rabbit breeds. Buck takes 18 months and doe 12 months for attaining its full size.

ARGENTE

The Argente breed has five varieties *viz.,* Argente de Champagne, Argente Creme, Argente Brun, Argente Blue and Argente Noir. The origin of this breed is France. Basically, it is a fur producing rabbit featuring the silver colour in the tip of undercoat along with black, blue, brown or orange colour guard hair depending on varieties.

The Argente de Champagne is oldest and largest in size among these varieties. This is a silvery rabbit of Champagne region of France. The breed has broad head, round skull, medium ears with round tips, dense, silky and glossy coat. The under coat is dark slate blue and body is bluish white evenly interspersed with long black guard hairs to give an 'old silver' effect. The adult body weight ranges from 3.5 to 4.5 kg.

The Argente Creme is the smallest among the Argente varieties. The under coat colour is orange and top colour is creamy white evenly and moderately interspersed with long orange hairs giving creamy effect. The eyes are brown in colour. Body weight is about 2 kg.

The Argente Brun, weighing about 2.5 kg, has short neck, cobby body, rounded short ears, broad and rounded loin and hind quarters. The under coat is deep brown and body colour is brownish white evenly interspersed with long dark brown guard hairs. They have very dense, glossy and silky coats in the body. The colour of eye is brown. It is very rarely seen now.

The Argente Blue has similar body structure to the Brun. But, the under coat colour is lavender blue with body colour bluish white, moderately and evenly interspersed with longer dark blue hairs. The eyes are also blue in colour.

The Argente Noir is also similar to the Brun . The under coat colour is deep slate blue and body colour is greyish white with evenly interspersed with longer black hairs giving silver effect.

BELGIAN HARE

The Belgian hare is a rabbit breed. As it has similar looks to the hare, sometimes it is confused with hares. The Belgian hare is first found in Flanders of Belgium during 1860. It is thought to be developed from Patagonian/Flemish Giant and Steenkonijn (Stone rabbit) rabbit breeds. They have long, slender body, large straight ears and long back legs and hence, are called 'Race Horse of the rabbit Fancy'. The common body colour is red but clack, tan and ruby eyed white are also occasionally seen. The body weight is around 4 kg. The litter size may vary from 5–12. They are extremely good pets.

BEVERANS

The Beveren breed is originated in the town of Beveren in western Belgium. The colour of original Beveren was lavender blue with a white face and white boots. Pure blue Beveren was developed through crossing. At that time blue variety was called Beveren and the blue with white rabbits were called the St Nocholas. After that, black, white, brown, lilac varieties were also developed. The Blue Beveren comes in a light lavender blue with blue-grey eyes. The Black Beveren has a glossy black colour with a dark blue undercoat and dark brown eyes. The White Beveren comes in pure white with brilliant blue eyes. They have mandolin shaped body and are used for both meat and fur production. The standard body weight is 3.5–5 kg. Does have good mothering ability and produce large litters. This breed has very fast growth rate.

CALIFORNIAN

During 1920's, Himalayan and Standard Chinchilla was crossed to produce a Chinchilla coloured buck. The buck was bred to New Zealand White does and the resulted breed was called Californian. Sometimes, it is confused with Himalayan, but both are different. This is a very good meat producing rabbit weighing about 3.5–4.5 kg. They have well developed head, short neck, proportionate ear carried upright position and straight tail. A dense undercoat with thick coarse guard hairs, pure white body, pink coloured bright eyes, black, chocolate, blue and lilac coloured nose, feet, ears and tail are the typical characters of this breed.

Chinchilla

The coat is very soft, grey and light blue in colour and resembles real chinchilla in texture. The chin and flanks are lighter in colour than the rest of the body. It's back and ears are laced with black colour whereas the belly and underside of the body upto the tail are lighter in colour.

French

The French Chinchilla is originated from France and is considered to be one of the best fur producing rabbits. It is believed that they are the descendent of crossing between Blue Beverans and Himalayan.

Soviet

The Soviet Chinchilla is originated from Soviet Russia and is considered to be one of the best meat producing rabbits. This breed has a very first growth rate, hence, is also called broiler rabbit. They attain market weight (2.0–2.5 kg) in 12 weeks of age. The adult body weight ranges from 3–5 kg. They are very prolific. The litter size varies from 7–12 or even more.

DUTCH

This attractive, good looking, fancy rabbit breed was originated in Netherland. They are excellent pet and show breed. The breed has typical marking on its body which consists of a white wedge down the face (the blaze), round coloured circles around the eyes and extending to the whisker bed (the cheek marking), white on the ends of hind feet (the stops), a thin white line extends between the ears (the hairline), a wedge shaped white marking on the back of the neck (the neck marking) and the contrast line dividing the coloured section of the body from the white portion (the top part- saddle and under the belly-undercut). The breed may be of lack, blue, chocolate, gray, tortoise and steel colour with mature body weight of about 2.5 kg.

FLEMISH GIANT

This is one of the oldest recognised domestic breed and is largest among all rabbit breeds in the world. It is originated in the province of Flanders in Belgium. The mature body weight varies from 6 to 10 kg or sometimes more. The most common colours are light grey or dark steel grey. Black, blue, sandy, fawn and white colours are also found. This fancy breed is kept for show purpose or as pet animals.

HARLEQUIN

The Harlequin rabbit breed was originated in France. The other varieties of this breed are Japanese-orange with coloured markings or Magpie-White with coloured markings. They have medium size body, broad head, hazel eye, erect 'V' shaped ears of different colours with the opposite side of their face being the same colour as the opposite ear. Colour of feet is also alternate. The body has short, soft, dense coat and is usually black and orange in colour but may also be of blue, black, brown, chocolate and lilac. The body weight is 3.0–3.5 kg. This excellent pet has the capability to produce 3–5 kits in one litter.

HAVANA

The Havana breed was first bred in Netherland in 1898. The original Havana was developed from the Dutch doe and unknown buck. They have short, rounded body with fairly short upright ears carried close together. The mature weight is 2–3 kg. Their coat is very glossy, soft and dense. The colour is dark and intense which may be of Chocolate, black and blue. They are very good companion rabbit.

HIMALAYAN

This breed is originated in the region of Himalayan Mountains. They are widely distributed throughout the world and have been known by different names *viz.,* Chinese, Russian, Egyptian, Black nose, Smut and Lapin. The breed has slender, cat like body, hence called 'Himalayan Cat'. Originally, Himalayan rabbit is white in colour with pink eyes and black nose, ears, tail and feet. The Blue, Chocolate and Lilac varieties are also created. The Himalayan rabbit is very famous for gentle temperament and is used as an excellent pet. They also produce elegant fur. The average body weight is around 2 kg.

LILAC

The Lilac is medium sized, docile, hardy and dove-blue coloured breed and should not be confused with the lilac clours of Rex, Netherland Dwarf etc. The Lilac rabbits are believed to be originated in a number of places within a few years. In 1913, Mr. H. Onslow of England showed this breed for the first time. In the same year, Miss Mabel Illingworth developed Lilac by crossing Blue Imperial with Havana. She named her new creation Essex Lavenders. Mr. C. H. Spruty of Holland created larger size rabbit called Gouda or Gowenaar in 1917 using Havana and Blue Beverans. In 1922, Prof. R. C. Punnet of Cambridge University created Cambridge Blue crossing Blue Beverans and Havanas. All three of the British strains were merged and the rabbit of the pinkish dove colouration became known as Lilac. It has number of strains which varies from different colours as well as sizes. This breed has slow maturity and smaller litter size (normally 4-6) but can be easily raised in all wire cages. The adult body weight varies from 2 or 3 kg.

LOP

This is one of the oldest domestic rabbit breed with extremely long ears about 20 inches hanging down the sides of the head and touch the floor. The breed is known as 'King of Fancy'. It is believed to be originated from England. Due to their high standard of excellence, during 1920's breeders refused to compete against any other variety class at shows and this breed was given a separate class of its own. The popular colour of English Lop is sooty fawn. Other colours are black, fawn and marked varieties. The body weight is about 4.5 kg.

The Netherland Dwarf is a smallest among all rabbit breeds weighing about one kg. The body size is slightly bigger than the guinea-pig. Their bodies and faces are round and ears are very short, no more than 5 cms long. They have big and bright eyes. The breed has a variety of colours including Agouti, Chinchilla, Squirrel, Lynx, Black, Blue, Brown, Lilac, White, Tan, Fox, Sable Marten, Fawn, Steel, Sable Siamese, Smoke Pearl Siamese, Tortoiseshell etc. The coat is short but soft. In spite of bad temperament of the buck, this breed is used as most popular pet due to their timidity and lively nature.

NEW ZEALAND WHITE

Generally, New Zealand rabbits are pure white with red eyes and pink ears. The brick red and black coat colours are also found. The New Zealand rabbit, originated in U.S.A., was first a red variety. Red variety is believed to be a cross between a Belgian Hare and Golden Fawn in 1900. The body is reddish white as underside is lighter but not white in colour. The white variety was derived from the cross of several breeds *i.e.*, Flemish or American White or Angoras and originated from the Shores of New Zealand. The white variety was first used for commercial meat and fur production. A black variety was also developed by crossing White buck and Red doe. The mature body weight reaches upto 5 kg.

PALOMINO

The Palomino breed was developed at the Lone Pine Rabbitry by Mark Young of Coulee Dam, in Washington State during the 1940's and early 1950's. This breed counts with medium body length with firm flesh, brown eyes, well developed shoulders, rounded and well filled hindquarters. The top line of the rabbit has a gradual arc rising upward from the neck to the loin and sloping downward towards the tail. The Palomino rabbit is a commercially used meat breed because of their fast growth rate and smaller bone compared to other meat rabbits. They are also known for being show pets due to their attractive appearance. This breed has two colour varieties- Golden and Lynx. The former variety comprises bright golden shade over a cream to white undercoat and the later, with medium pearl grey surface colour blending to an orange beige intermediate colour over a cream to white undercoat. The average body weight of full grown rabbit is 4.0 kg.

POLISH

The Polish rabbit, despite the name, is believed to have originated in Belgium. Probably this breed was developed from Dutch and Himalayan rabbits in the 1860's for the meat purpose. In the 1900's the Polish rabbit became one of the most popular meat breeds in Europe, especially in Belgium. Although they are one of the smallest breeds of rabbit but not actually dwarf breed. They have compact, elongated, and arched body, small and rounded head, long front leg and a wedge shaped face topped with short upright ears. The fur is short, dense with fine silky texture which appears as if it has been polished. The coat comes in variety of colours *viz.,* black, blue, agouti and albino white. This breed became more popular as a pet. The average body weight is about 1.0 kg.

REX

The Rex breed was first appeared in a litter of wild gray rabbits and was originated in France in 1919 as a result of a recessive genetic mutation leading to the shortening of guard hairs. In 1924, it was first introduced in rabbit show. This breed is commercially used for fur and meat purpose. The rex gene also occurs in cat where the cat is found to be completely bald but Rex rabbits have significantly thickened coat. The Standard Rex is a variety of rabbit recognised for its upstanding velvety textured fur. The fur length varies from 1.3 to 2.2 cm in length and the guard hairs are of equal length as the undercoat. Mini Rex is a smaller version of the Rex rabbit weighing about 2 kg. Combing is not frequently essential for these rabbits. The Standard Rex breed is found in variety of colours like black, orange, white, blue, chocolate, sable, tortoise, chinchilla, Californian, Siamese, seal, brindle, castor, lilac, lynx, broken, red, opal etc. It has broad head, upright ears and round medium size body weighing around 3.5 kg. The doe has large dewlap. High litter size and excellent foster mothering are the additional qualities of the female Rex breed.

RHINELANDER

The Rhinelander rabbit was developed in Germany in 1900 and was shown for the first time in 1902. The breed was imported to the United States in 1923. During the war years the breed disappeared. Then, they were again imported to United States in 1972. This is a distinct type of full arched breed having long leg and medium bone structure allowing the animal to move easily and graceful down the table. Their cylindrical body, similar width of shoulder, midsection and hip, smooth top line and well rounded but not heavy hind quarters, give them sleek appearance. It is a meat rabbit with a dense, silky coat, white body with black and bright golden orange spots and brown eyes. The markings on a Rhinelander are typical of the butterfly marked breeds. The head marking includes coloured ears, eye circles, cheek spots and the butterfly with no ragged edges. The spine marking is different from any of the other marked breeds. The edges of the spine marking are to be even. The spine starts narrow at the base of the neck, widens at the shoulder and then narrows again at the base of the tail. The side marking is ideally six to eight spots evenly distributed over the hip and loin. There is marking in each side of the body. A spot is a coloured area completely surrounded by white. A spot may be either colour or both, but, both colours must be present on each side. The full mature body weight is 3 to 4.5 kg.

SABLE

The American Sable is a cross of the Chinchilla rabbit breed with rounded body. The Sable variety has rich sepia brown on the body with darker sepia brown on the ears, face, back, legs, and upper portion of tail. The American Sable has its unique and distinct coat colouring and the ruby red glow in its eyes. The coat is medium in length, fine, soft and very thick. The full grown rabbit is 3–4.5 kg.

SATIN

The Satin breed was developed by Walter Huey from the defective litter of Havana rabbits. The colour varieties include black, blue, chocolate, Californian, lilac, chinchilla, copper red, Siamese, ruby-eyed white and broken group. They have broad head, short neck, proportionate ears, and well developed hind quarters with curved top line. The breed has translucent hair shafts that reflect light and give its coat a unique shine. The fur is dense and silky. The mature body weight is around 3.5 kg. Satin rabbits are used for showing, house pet and commercial meat production.

SIBERIAN

The Siberian rabbit was produced in U.K. in 1930. It is used as fur breed and should not be confused with Himalayan rabbit. They have translucent hair shaft, shinny coat of about one inch length and rounded hind quarters. The coat colour may be of black, blue, brown or lilac. The mature body weight is 4–5 kg.

Smoke Pearl

This breed was originally known as Smoke Beige. Although this breed was created in Scotland in 1926, it was renamed as Smoke Pearl in 1932. They have moderate body length, erect ears and arched back feet. The fur of this popular pet is dense, soft, one inch in length and smoky in colour with pearl grey on both the sides. The popular colour is light pearl grey and beige. The adult body weight varies from 2 to 3 kg.

THE FOX

The silver fox breed is believed to be originated and developed during 1920's in U.S.A. In 1925, the breed was accepted as 'American Silver Giant' and in 1929, this was renamed as 'American Silver Fox'. This breed is not recognized in other countries, although in Australia, New Zealand and the United Kingdom there is a breed called Silver Fox which is actually the same rabbit breed as the Silver Marten in America. The Silver Fox has black and blue varieties. It is a multipurpose breed, reared for meat and fur production. The mature body weight of Silver Fox is 3.5 to 5 kg. The body has well developed muscle in shoulders, over the top, ribs and hind quarters. The fur is unusually long, coarse and coat is evenly silver in colour which resembles the pelt of a fox. When the fur strokes from tail to head, it stands straight up until stroked in the opposite direction. Head, flanks, tail and feet have silver tipped or white guard hairs. This breed has very high *i.e.*, 65% dressing percentage compared to other rabbit breeds. They are easy to breed, highly adaptable to variable climate and are famous for their very good mothering ability.

THRIANTA

The Thrianta rabbit was developed in Netherland from Black Tan, English Spot and Havana in 1940. During World

War II, just days after the breed's first official recognition, the German forces invaded the Netherlands and the total population of this breed was affected. The resurgence of this breed occurred during 1970's when it became accepted in British Rabbit Council Standard and a special club dedicated to the breed was formed in Netherlands. It was introduced in U.S.A. in 1996. This breed is known as 'Fire of the Fancy' due to its brilliant red orange coat and brown eyes. The Thrianta has medium size body weighing about 2.5 kg. Today, they are used as a very popular show and pet rabbits.

VIENNA

The Vienna rabbit is an old breed of Austria. They appear in two colour varieties-the Vienna Blue and the Vienna White. The former has a dark, slate blue top colour with a lighter under colour and the later, has a silky white coat. Their fur is very soft. The mature body weight ranges between 3.5 to 4.5 kg.

Chapter 3

Nutrition and Feeding Management

Rabbits are herbivorous rodents. They eat fresh green plants, grasses, garden greenery, outer leaves of cabbages, cauliflowers, tops and bottoms of carrots and other left over salad materials including kitchen plants. The wild rabbit browses on grass and prefers leguminous plants. In case of captive rabbits, there should be regular supply of fresh green forages in their ration. In addition, commercial rabbits require '*balanced concentrate ration*' to meet out their nutrient requirements for optimum body growth, production, reproduction and health. The concentrate mixture includes wide variety of grains *e.g.*, maize, oat, barley etc. as energy source, cakes *e.g.*, soyabean meal, ground nut cake, mustard cake, sesame cake, til cake etc. as protein source, and salt, vitamins and minerals. A large variety of ingredients are recommended to use in the mixture to fulfil the deficiency of different nutrients and also to change the palatability of feed. The concentrate feed should be given in pellet form to avoid dustiness of feed. Dusty feed can cause respiratory problems in rabbits. Moreover, the teeth of rabbits are continuously growing in nature throughout life. To maintain their teeth length, the diet should be included hard vegetables like carrot, turnip etc. and pellet feed. Rabbit prefers the feed pellet of 3–4 mm in diameter and 10–15 mm in length. The feeding time must be fixed in a daily routine to avoid any stress to them. Normally, the concentrate ration is given in the morning after cleaning the feeding trough and green fodder once in the afternoon. Some cases, green fodder is given twice daily once in the morning and once in the afternoon.

DIGESTION IN RABBIT

The functioning of rabbit digestive tract is similar to that of other monogastric animals. Digestion takes place in many stages in the digestive tract. It comprises breaking down of complex food into smaller particles which the rabbit can readily absorb into the body. This breaking down of food can be taken place in the presence of enzymes and special chemicals, secreted from various part of digestive system. As a result, protein turns into amino acids, carbohydrates into sugars, and fats into fatty acids and glycerides.

The whole food materials are cut into small pieces by incisor teeth and grinded by check teeth. In the mouth, it is mixed with saliva, secreted by salivary glands. It reaches into stomach in its finely grinded state and mixes with acidic fluid and enzymes. The food is stored in this form near pylorus for further digestion. When they pass into the small intestine numbers of enzymes are mixed with it in addition to bile salt from liver. The bile acts upon the fat of the food and with other enzymes it is broken into fatty acids and glycerides. In small intestine the remaining part of food is also finally digested during its passage and absorbed through villi into blood system. The digested fat portion is absorbed into lymphatic system. Up-to 90% of the protein, starches and sugars are absorbed from the food in the small intestine.

The undigested fibrous part of food materials, generally larger in size, further moves on, sorted and goes to colon forming a hard waste. While the remaining food particles, generally smaller in size, is ready for digestion and goes to caecum. The pellets are mixed with mucous and become soft in caecum. Then, the rabbit twists their own body around, sucks the soft faeces as they emerge from the anus and swallows intact without chewing them. These soft pellets also follow the same digestive process as normal feed materials. This feeding habit of rabbit is known as *coprophagy* or *caecophagy*. In domestic rabbit, it is practised at night/early morning while in wild rabbits it occurred at day time during the stay in burrow. Soft pellet contains twice the protein, half the fibre of the hard pellet and other essential nutrients. However, coprophagy improves nitrogen reutilization through the re-ingestion of bacterial protein synthesis in the caecum. The final residue of the twice digested food passes into the colon in the form of hard pellet and excreted through rectum as faeces. Thus, the nutrients, obtained from digestion, are transported to various parts of the body through bloodstream and surpluses, are stored in various places in different forms for future use.

The composition of hard and soft faeces of rabbits fed different feeds are presented in Table 5.

FEED RESOURCES FOR RABBITS

There are two main feed resources: (*a*) roughages, and (*b*) concentrates/pellets which are used for feeding rabbits.

Roughage

Roughage contains high fibre and low digestible energy compared to grains. This is bulky in nature and includes dry as well as green roughages. Both dry and green roughages must be incorporated in rabbit diet to fulfil the fibre and other valuable nutrient requirements in the body. Fibre plays vital role for the normal functioning of the digestive system in rabbit. It prevents diarrhoea and to form hair balls in the stomach. Paddy straw, wheat straw, dry

Table 5: Composition of hard and soft faeces (based on balance concentrate feed, green and dry forages and ten different feeds)

Components	***Hard pellets***		***Soft pellets***	
	Average	***Range***	***Average***	***Range***
		Percentage		
Moisture	41.7	34–52	72.9	63–82
Dry matter	58.3	48–66	27.1	18–37
		Percentage of dry matter		
Proteins	13.1	9–25	29.5	21–37
Crude fibre	37.8	22–54	22.0	14–33
Fats	2.6	1.3–5.3	2.4	1.0–4.6
Minerals	8.9	3.1–14.4	10.8	6.4–10.8
Nitrogen-free extract	37.7	28–49	35.1	29–43

Source: Proto (1980)

grasses, hay etc. are the sources of dry roughage. Whereas, leguminous green crops and forages *e.g.,* lucerne, berseem, cowpea, soyabean, clovers, ground nut, horsegram etc. and non-leguminous green fodders *e.g.,* maize, oat, barley, jowar, different green grasses etc. harvested at pre-flowering stage are the sources of green roughages. Rabbit prefers leguminous green fodders. In addition, they should also be provided different root crops *e.g.,* turnip, carrot, radish, beet, sweet potato etc. to prevent overgrowth of teeth. They are relish eater of cabbage, cauliflower, broccoli, spinach, lettuce etc. The leguminous green roughages provide 13–20% and non-legumes 5–10% crude protein in the diet. The dry matter content depends upon type of forages, weather, stage of harvesting, season etc. Normally, straw supplies approximately 90% DM, hay 75–80% and green roughage 5–20% in rabbit diet. A variety of roughage must be included daily to ensure proper intake.

Concentrate

Concentrate has high protein (more than 18% CP), high digestible energy and low fibre (less than 18% CF). The dry matter content of concentrate ration should be around 90%. In rabbit, concentrate mixture is fed in the form of pellet because: (*a*) it avoids respiratory diseases as the dust particles of mesh feed are inserted into rabbit nostrils during feeding, (*b*) ingredients are properly mixed, (*c*) wastage is minimum, (*d*) improves palatability resulting more consumption and high feed conversion efficiency, and (*e*) maintains teeth length. Efforts are made to use maximum numbers of ingredients from different sources to avail all the nutrients in rabbit pellet. The ingredients, used in feed, must be fresh and free from any deleterious microbes and toxic substances. It should be used within three months of preparation or manufacture to avoid aflatoxin or any spoilage. Cereal grains like maize, oat, barley, sorghum, wheat and byproducts like wheat bran, rice bran, molasses, fruit pomaces, beet pulps etc. are used as energy source. On the other hand, cakes/meals of ground nut, soyabean, sunflower, til, rapeseed, linseed, peanut, grams, fish, meat, feather, silk worm etc. are the different protein source feed ingredients. To promote the growth and to improve feed intake of rabbit, feed additives *viz.,* probiotics, antibiotics, buffers, organic acids can be mixed in pellet. Furazolidone @5 gm per quintal or sulphaquinoxaline @0.025% is commonly used in commercial rabbit pellet feed. The rabbit pellet is generally prepared from 65–67% energy ingredients, 30–32% protein ingredients, 1.8–2.0% minerals, 0.8–1% salt and 0.2% vitamins. If required, feed additives may be added. The pellet should be neither too hard nor too fragile because both cause low intake.

Beside these, sometimes rabbits may be offered 1–2 tablespoons of fresh fruits in their daily diet as '*treats*'. But, high carbohydrate (starchy) or fat treats *viz.,* peas, beans, bread, breakfast cereals, biscuits, nuts, seeds, crisps, chocolate, honey sticks etc. must be avoided as it can cause digestive disorders.

The chemical composition of different feed ingredients suitable for rabbit feed is shown in Table 6.

The recommended incorporation limits of various minerals, vitamins and selected amino acids in rabbit feed is given Table 7.

NUTRIENT REQUIREMENTS

Rabbits are used for mainly meat, wool and fur production. In addition, they are also kept as laboratory experimental animals, pet animals and to take part in animal shows. The special care is needed in feeding to fulfil the nutrient requirements of these animals when they are particularly managed for commercial meat, wool or fur production. However, the nutrient requirements may vary according to breed, sex, age, ambient temperature, kind of feed, breed utility, state of life *i.e.,* breeding, pregnancy, lactation (Table 8) etc. The requirements of basic nutrients are described briefly as under.

Carbohydrate

Carbohydrates are important energy source. Additional carbohydrates are converted into fat and are broken down as and when required by the body. Carbohydrate can be classified according to their structural complexity *viz.,* monosaccharides, oligosaccharides, polysaccharides and complex carbohydrate (glycoproteins). It can also be categorized into sugar, starch and fibre.

Like other species, simple monosaccharide sugars *e.g.,* glucose, fructose and galactose are absorbed from the small intestine. Polysaccharides or starches are abundant in seeds,

Table 6: Chemical composition of different feed ingredients used in rabbit feed

Feed	*DM*	*F*	*CF*	*ICF*	*TN*	*Lys*	*SAA*	*Mx*	*Ca*	*P*	*DE*
Oats	86	5.3	10.2	9.8	10.0	0.40	0.50	2.70	0.08	0.34	2800
Wheat	86	1.9	2.3	1.0	11.3	0.32	0.47	1.65	0.06	0.33	3100
Maize	86	4.2	2.2	0.6	9.0	0.25	0.39	1.35	0.01	0.27	3200
Barley	86	2.0	4.0	3.8	10.0	0.37	0.42	2.30	0.05	0.35	3000
Sorghum	86	3.0	2.5	1.0	10.0	0.23	0.33	1.45	0.03	0.30	3150
Paddy rice	87	2.1	8.6	6.5	8.0	0.28	0.35	4.53	0.05	0.26	2850
Fine wheat bran	87	4.0	9.6	6.8	15.0	0.56	0.50	5.60	0.13	0.20	2300
Wheat flour (II quality)	88	2.7	1.4	0.10	14.9	0.50	0.46	2.00	0.07	0.45	3200
Brewery draff	91	7.6	15.3	3.5	25.2	0.70	0.61	4.07	0.28	0.50	2800
Maize bran	89	6.3	9.0	3.8	10.1	0.27	0.36	2.69	0.03	0.23	2750
Soya cake 44	88	1.8	7.4	6.8	42.5	2.70	1.27	6.00	0.30	0.62	3260
Soya cake 45	88	2.0	5.6	4.8	45.8	2.91	1.37	6.30	0.30	0.69	3310
Sunflower cake	90	1.8	26.5	18.6	29.5	1.07	1.26	6.22	0.35	0.90	2770
Rapeseed cake	89	1.8	11.7	7.4	35.2	1.93	1.73	7.00	0.75	1.10	2800
Cottonseed cake	91	1.4	13.0	9.0	41.0	1.72	0.59	6.46	0.20	1.00	2790
Horse bean	87	1.30	7.5	5.0	26.4	1.66	0.53	3.38	0.11	0.61	2800
Field pea	86	1.60	5.5	4.0	22.0	1.6	0.59	3.40	0.08	0.45	2800
Extruded soya	89	18.0	6.0	4.2	37.0	2.35	1.15	4.45	0.25	0.57	4400
Grass meal	91	3.7	21.0	14.3	17.1	0.75	0.44	12.7	0.70	0.42	1730

Table 6: (*Contd...*)

Table 6: (*Contd...*)

Feed	*DM*	*F*	*CF*	*ICF*	*TN*	*Lys*	*SAA*	*Mx*	*Ca*	*P*	*DE*
Dehydrated alfalfa A	90	3.0	27.0	22.0	15.5	0.68	0.42	9.0	1.40	0.25	1800
Dehydrated alfalfa B	90	2.90	25.0	22.5	16.6	0.73	0.45	9.45	1.50	0.25	1850
Soyabean husks	92	2.0	34.0	32.0	12.7	0.70	0.35	5.69	0.40	0.17	1800
Cocoa husks	90	4.5	18.6	14.0	16.5	0.90	0.38	7.62	0.30	0.35	2190
Wheat straw	88	1.3	42.0	39.0	4.0	0.20	0.12	8.30	0.47	0.09	700
Sugar-beet pulp	90	1.0	18.0	5.0	8.8	0.54	0.13	5.42	0.90	0.11	2700
Citrus pulp	90	3.0	12.0	5.1	6.0	0.25	0.06	5.45	2.10	0.12	3000
Gluten feed	90	3.0	8.3	4.6	21.0	0.69	0.97	7.10	0.28	0.70	2770
Cassava	85	0.17	4.6	2.0	2.6	0.09	0.06	5.22	0.30	0.19	2850
Carob bean	86	2.4	7.8	7.0	5.0	.18	.16	3.43	0.65	0.10	2390
Sugar-beat molasses	77	0.3	0.0	0.0	7.7	0.04	0.10	8.93	0.25	0.02	2600
Animal fat	99.5	99.5	0.0	0.0	0.0	0.0	0.0	0.0	0.0	0.0	8000
Soyabean oil	99.5	99.5	0.0	0.0	0.0	0.0	0.0	0.0	0.0	0.0	8500
Meat meal A	92	7.5	0.0	0.0	59.0	3.46	1.39	22.7	7.05	3.35	3180
Meat meal B	95	14.5	0.0	0.0	58.2	3.40	1.34	19.3	6.55	3.10	3680
Fishmeal	91	8.3	0.0	0.0	67.8	5.00	2.50	15.0	3.90	2.55	4160

DM= Dry matter, F = Fat, CF = Crude fibre, ICF= Indigestible crude fibre, TN = Total nitrogen, Lys = Lysine, SAA = Sulphurous amino acids, Mx = Total minerals, Ca = Calcium and P = Phosphorus as percentage of feed. DE = Digestible energy in kcal/kg of feed.
Source: INRA (1989) and Maertens *et al*. (1990)

Table 7: Recommended incorporation limits of various minerals, vitamins and selected amino acids in rabbit feed

	Deficit	***Observed minimum with no problems***	***Optimum***	***Observed maximum with no problems***	***Toxicity symptoms***	***Stage***
Minerals *(ppm)*						
Calcium	700	3000	4000	25000	40000	Growth
	3000	8000	12000	19000	25000	Reproduction
Phosphorus	1 200	2600	3000	8000	–	Growth
	4000	4500	6000	8000	10000	Reproduction
Sodium	–	2000	3000	6000	7000	Growth
Potassium	3000	6000	6000	16000	–	Growth
	–	–	9000	16000	20000	Reproduction
Chlorine	1 700	2500	3200	4200	–	Growth
Magnesium	200	–	2500	3500	4200	Growth
Manganese	–	–	8.5	–	50	Growth
	0.6	–	13.0	–	–	Reproduction
Iodine	–	–	0.2	–	10000	Growth

Table 7: *(Contd...)*

Table 7: (*Contd...*)

	Deficit	***Observed minimum with no problems***	***Optimum***	***Observed maximum with no problems***	***Toxicity symptoms***	***Stage***
	–	–	0.2	–	100	Gestation
Fluoride	–	–	0.5	–	400	Growth
Copper	2	3	5	150–200	200–300	Growth
Zinc	2	7	50	85	–	Growth
Vitamins *(/kg)*						
Vitamin A *(IU)*	–	3000	10000	20000	75000	Reproduction
Vitamin D *(IU)*	–	600	1000	2000	3000	Reproduction
Vitamin E *(mg)*	17	–	50	–	–	Growth
	17	25	50	–	–	Reproduction
Amino acids *(g/16 gN)*						
Lysine	2.50	3.75	4.40	7.5	9.4	Growth
Sulphur AA	2.50	3.00	3.75	4.4	5.0	Growth
Arginine	3.00	3.75	5.60	12.5	–	Growth
Tryptophane	–	0.75	0.80	1.60	–	Growth

Source: FAO (2010)

Table 8: Nutrient requirements of rabbits at different stages fed *ad libitum* (percentage or amount per kg of diet)

Nutrients	***Maintenance***	***Growth***	***Gestation***	***Lactation***
Digestible energy (kcal/kg)	2100	2500	2500	2500
Total digestible nutrients (%)	55	65	58	70
Crude protein (%)	12	16	15	17
Crude fibre (%)	14	10–12	10–12	10–12
Fat (%)	2	2	2	2

Source: Nutrient requirements of rabbits, second revised edition, 1977 (Subcommittee on Rabbit Nutrition Committee on Animal Nutrition Board of Agriculture and Renewable Resources National Research Council)

fruits, tubers and roots and are broken down into simple sugars during digestion. Plant originated starches are present as granules and insoluble in cold water. When a suspension of starch in water is heated, the granules swell and gelatinize. Cereal starches are more fermentable than tuber or root originated starches. The reaction is catalysed by amylase which is secreted by the salivary glands and pancreas and is also present in caecotrophs resulted from bacterial synthesis. The type of starch, dietary level and age of rabbit influence digestion and absorption in the small intestine. Microbes in rabbits produce more volatile fatty acids on starch based diet than on forage diets. However, regular feeding of starchy and sugary treats to rabbit can cause obesity and enterotoxemia. Too fine grains also lead to rapid fermentation of the starch causing enterotoxemia. To avoid this, coarse grind feed is recommended in the diet. The dietary fibre plays crucial role in maintaining gut health, stimulating gut motility, reducing fur chewing and preventing enteritis in rabbit. The requirement is quite higher in rabbit compared to other species. They need minimum 20–25% fibre in the diet depending upon different categories. Not only that, the fibre particle size also affects the normal functioning of caecum as well as colon. Digestible fibres are fermented in caecum converting into volatile fatty acids and indigestible fibres are expelled out from the body. They can be digested and absorbed from the stomach and small intestine or degraded and fermented by the caecal micro flora. The deficiency causes low growth rate, diarrhoea and 'fur eating' habits.

Protein

The role of dietary protein in animal body is well recognized. The protein, composed of essential and non-essential amino acids, is important in rabbit diet for optimum growth, production and reproduction. Leguminous plants, vegetables and concentrate mixture etc. are the different protein sources. The essential amino acids are not synthesized by animal's body but must be supplied through diet. There are also certain amino acids which can be replaced by other *e.g.*, methionine by cystine, tyrosine partly by phenylalanine. In rabbits, the essential amino acids are arginine, glycine, histidine, isoleucine, leucine, lysine, methionine or cystine, phenylalanine or tyrosine, threonine, tryptophan and valine. Due to the nature of coprophagy, the essential amino acid requirement of rabbit is quite complicated. The caecal microorganisms synthesize some essential amino acids especially lysine, sulphur containing amino acids, and threonine. The composition of amino acid depends upon microbial population and dietary protein digestibility. The research results have shown that microbial protein plays a minor role in meeting a rabbit's protein or amino acid needs. The majority of microbial protein utilized by rabbit is digested in the colon. Therefore, synthetic amino acids particularly lysine and metheonine must be supplemented through rabbit diet to meet out amino acid requirements. By nature, rabbits are very efficient to utilize forage protein. Addition of forage source in the diet may be the alternate for this.

The level of protein in the diet depends on growth, pregnancy, lactation etc. Growing rabbits require more protein than other categories as the growth rate at this stage is very fast. The dietary protein and energy influence the ingestion of caecotrophs. Rabbits fed low dietary protein enhance caecotroph consumption and high level protein decreases consumption which may be due to the protein sparing mechanism of the body. Utmost care should be taken when feeding high level dietary protein as it causes respiratory diseases due to excess ammonia secretion, eye infections, and '*sticky bottom syndrome*' due to incomplete coprophagy. High caecal ammonia level also increases caecal pH which allows pathogens to flourish in the gut leading to enteritis.

Fat

Fat is also an energy source of the diet. It is easily digested in the small intestine. Fat contains 2.25 times more energy per weight as carbohydrate and it is a much more efficient means to store energy in the body. Lots of energy can be supplied with little amounts added through diet. Fat generally increases the absorption of fat soluble nutrients, such as fat soluble vitamins. The requirement of dietary fat of rabbits has been set at 2% by many researchers which can be fulfilled from the rabbit pellet feed available at the market. Being a basic consumer of plant feedstuffs, rabbit satisfies it's all essential fatty acid requirements and maximizes absorption of fat soluble vitamins at this level. But, fat levels in feed as high as 25% have no ill effects on the rabbit. Rather, it improves palatability of poor quality feeds and feed conversion efficiency. Rabbit keepers use to add little oil to feed top dressings in order to improve rabbit coat condition. A high fat level is not generally recommended in hot environment probably because of high temperature and high cost of supplementary fat.

Water

Water is one of the vital nutrients of rabbit diet. Ideally, rabbit should have access to an unlimited source of fresh and clean drinking water at all times. Water may be provided either in bowls or drop feeders or bottles with stainless steel lids. Generally, bowl is avoided because: (*a*) it develops skin infection as dewlap of rabbit becomes continuously wet during drinking, and (*b*) water bowls are prone to contamination with dirts, hairs/fur, faeces and urine. Assure that water flow is unobstructed and free-flowing when touched by rabbit. Change the drinking water daily and sanitize water bottle/bowl with mild dish detergent and rinse thoroughly on weekly basis. On an average, one rabbit consumes 50–150 ml of water per kg of body weight per day. Water intake depends on environmental temperature, kind of feed (DM content), age, sex, breed, stage of pregnancy and lactation. They even drink twice as much water during hot environment (30°C or more) as they take during cold climate (10°C or less). High fibre diets require extra water to moisten the feed and to maintain the adequate fluid level in the digestive tract. High protein diet also requires more water to excrete excess urea through urine as kidney has a limited capacity to concentrate excretory by products in the urine. Hot water during summer and cold water during winter affect the quantity of intake. Water quality is an important factor for its consumption. Poor quality *e.g.,* foul odour, bad tastes, slimed water due to dust, spilled and faecal matter contamination may discourage rabbits from drinking. It affects health and sometimes, leads to death of animals. Most surprising thing is that when rabbit does not have access to feed, it drinks more water. After three days food deprivation, one rabbit even consumes about six and half times of its normal intake.

On the other hand, if they are deprived of water for three days at a stretch, they will stop eating entirely. Rabbit can't tolerate water deprivation for more than 24 hours during hot weather.

Vitamins

Dietary vitamin requirements in rabbits are quite low. They require essential vitamins, or their precursors, to be present in the ration. Vitamins are normally present in fresh foods offered to the rabbit and the deficiency is rare. Some of the vitamins like vitamin B complex, C and K are synthesised by the caecal micro flora during digestion. So, they do not have to be present in the ration. Vitamins A, D and E are to be included in the concentrate ration. Concentrate should not be stored for more than three months as the vitamin content deteriorates gradually over the time. However, the requirements of different vitamins are described as under.

Vitamin A

Vitamin A or retinol is a fat soluble, organic alcohol formed in the intestinal mucosa of rabbit primarily from β-carotene of plants. It is stored in the liver and is circulated in the body according to the need. Carrots, colour fruits, fresh green vegetable and plants are the good sources. Vitamin A helps in normal vision, maintenance of epithelial cell integrity, reproduction, bone development, immunological responses and body growth. The dietary requirement of vitamin A is 10000–18000 IU/kg. The recommended incorporation limits of various minerals, vitamins and selected amino acids in rabbit feed is given in Table 9.

Vitamin D

Vitamin D, a fat soluble vitamin and also a hormone, plays important role in calcium and phosphorus metabolism. In

Table 9: Recommended chemical composition of feeds for intensively reared rabbits of different categories

Components of feed, assumed to contain 89 percent dry matter	***Unit***	***Growing rabbits (4–12 weeks)***	***Lactating doe + young under-mother***	***Pregnant doe, not lactating***	***Resting adults (males)***	***Mixed breeding does plus fatteners***
Crude proteins	%	16	18	16	13	17
Amino acids						
Methionine + cystine	%	0.60	0.60	–	–	0.60
Lysine	%	0.65	0.75	–	–	0.70
Arginine	%	0.90	0.80	–	–	0.90
Threonine	%	0.55	0.70	–	–	0.60
Tryptophane	%	0.18	0.22	–	–	0.20
Histidine	%	0.35	0.43	–	–	0.40
Isoleucine	%	0.60	0.70	–	–	0.65
Phenylalanine + tyrosine	%	1.20	1.40	–	–	1.25
Valine	%	0.70	0.85	–	–	0.80
Leucine	%	1.05	1.25	–	–	1.20
Crude fibre	%	14	12	14	15-16	14
Indigestible crude fibre	%	12	10	12	13	12
Digestible energy	kcal/kg	2500	2600	2500	2200	2500

Table 9: (*Contd...*)

Table 9: (*Contd...*)

Components of feed, assumed to contain 89 percent dry matter	***Unit***	***Growing rabbits (4–12 weeks)***	***Lactating doe + young under-mother***	***Pregnant doe, not lactating***	***Resting adults (males)***	***Mixed breeding does plus fatteners***
Metabolizable energy	kcal/kg	2400	2500	2400	2120	2410
Fats	%	3	3	3	3	3
Minerals						
Calcium	%	0.40	1.10	0.80	0.40	1.10
Phosphorus	%	0.30	0.80	0.50	0.30	0.80
Potassium	%	0.60	0.90	0.90	–	0.90
Sodium	%	0.30	0.30	0.30	–	0.30
Chlorine	%	0.30	0.30	0.30	–	0.30
Magnesium	%	0.03	0.04	0.04	–	0.04
Sulphur	%	0.04	–	–	–	0.04
Cobalt	ppm	0.1	0.1	–	–	0.1
Copper	ppm	5	5	–	–	5
Zinc	ppm	50	70	70	–	70
Iron	ppm	50	100	50	50	100
Manganese	ppm	8.5	2.5	2.5	2.5	8.5
Iodine	ppm	0.2	0.2	0.2	0.2	0.2

Table 9: (*Contd...*)

Table 9: (*Contd...*)

Components of feed, assumed to contain 89 percent dry matter	***Unit***	***Growing rabbits (4–12 weeks)***	***Lactating doe + young under-mother***	***Pregnant doe, not lactating***	***Resting adults (males)***	***Mixed breeding does plus fatteners***
Fluorine	ppm	0.5	–	–	–	0.5
Vitamins						
Vitamin A	UI/kg	6000	12000	12000	6000	10000
Vitamin D	UI/kg	900	900	900	900	900
Vitamin E	ppm	50	50	50	50	50
Vitamin K	ppm	0	2	2	0	2
Vitamin C	ppm	0	0	0	0	0
Vitamin B_1	ppm	2	–	0	0	2
Vitamin B_2	ppm	6	–	0	0	4
Vitamin B_6	ppm	2	–	0	0	2
Vitamin B_{12}	ppm	0.01	0	0	0	0.01
Folic acid	ppm	5	–	0	0	5
Pantothenic acid	ppm	20	–	0	0	20
Niacin	ppm	50	–	–	–	50
Biotin	ppm	0.2	-	-	-	0.2

Source: Lebas (1979) and Lang (1981)

the presence of ultraviolet rays, endogenous vitamin D precursor, 7-dehydrocholesterol, converts to pre-vitamin D_3 in the skin. Before it is transported to liver, it binds with plasma protein and converted into vitamin D_3 or cholecalciferol. In plants, ergosterol is present as vitamin D precursor which ultimately converted into ergocalciferol or vitamin D_2. This takes place during the drying of plants in the sunshine. Then, it is hydroxylated in the liver and converted to active vitamin D in the kidney in the presence of parathyroid hormone. Deficiency of vitamin D causes slow absorption of phosphorus and calcium resulting hypophosphatemia, osteomalacia and rickets in rabbits. Rabbit diet should contain 1000 IU/kg of vitamin D. Dietary level above 2300 IU/kg causes toxicity. It mineralizes the soft tissues specially in kidneys and aorta. As vitamin D is rare in foods, it is advised to expose rabbits to sunlight for vitamin D synthesis in their body. In most species, vitamin D is required for active transport of calcium from the gut to the bloodstream but in rabbit with sufficient dietary concentration calcium is absorbed efficiently and passively from the gut even in the absence of vitamin C. Passive calcium absorption is dependent on a diffusion gradient across the gut wall. Low dietary calcium results in lower gradient and less absorption.

The ratio of vitamin D to vitamin A is crucial as they interact with each other and work like hormone. These vitamins are fat soluble and are stored in liver for future use. Rabbit is an efficient converter of vitamin A from beta-carotene. Further, addition of synthetic vitamin A and D in the diet can cause toxicity in animals. Therefore, care is needed during supplementation of these two vitamins in rabbit diet.

Vitamin E

Vitamin E, a fat soluble vitamin and natural antioxidant, acts synergistically with selenium and inactivates the peroxides which cause widespread tissue damage. Green forage and cereals are very good source of natural vitamin E. Immature plant leaves and germinated grains have very high level of vitamin E. In rabbit, liver infection caused by *Eimeria steidae* affects the metabolism of fat soluble vitamins causing muscular dystrophy. Deficiency also causes reproduction and health related problems. The requirement of vitamin E in rabbit is about 40 mg/kg which increases with dietary level of polyunsaturated fatty acids and vegetable oils.

Vitamin K

This is a fat soluble vitamin and is known as blood clotting factor. Sufficient amount of vitamin K is synthesized by the microbes in the rabbit gut to meet out the normal body requirements. Green grasses are very rich source of vitamin K. Green fed rabbits rarely suffer from the deficiency of this vitamin. If green fodder is not available, a level of 2 ppm of synthetic vitamin K in concentrate is sufficient to prevent haemorrhage and abortion in rabbits.

Vitamin B complex

Vitamin B complexes are water soluble. The requirements of various members of vitamin B complexes are either partially or even completely fulfilled through the routine practice of coprophagy. Vitamin B complexes including thiamine (Vit.B_1), riboflavin (Vit.B_2), niacin or niacinamide (Vit.B_3), pantothenic acid (Vit.B_5), pyridoxine or pyridoxamine (Vit.B_6), choline and B_{12} or cyanocobalamin are synthesized in the intestinal tract, are recycled through caecotrophs and are made available to the rabbits. Prolong feeding of thiamine deficient diet caused mild ataxia in rabbit. The dietary deficiency of niacin has resulted in pronounced loss of appetite followed by emaciation and diarrhoea. A significant growth in rabbit can be achieved by feeding additional niacin up to 11 mg/kg of body weight. However, cobalt supplementation is required in diet to synthesize vitamin B_{12}. Research results show that rabbits get benefitted with feed efficiency and growth by adding 1300 mg/kg choline, 1 mg/kg pyridoxine, 10 mg/kg cyanocobalamin and 50 mg/kg nicotinic acid in their diet. These vitamins do not cause any toxicity as excess is excreted through urine.

Vitamin C

Vitamin C or ascorbic acid is a water soluble vitamin. It is synthesized from glucose in the liver and helps in maintaining as well as repairing of connective tissue. In rabbit, dietary supplementation is not generally required. But, 50–100 mg/kg daily can be used to treat respiratory diseases and to prevent enterotoxemia by inhibiting toxin production. Overdosage of vitamin C is not harmful to rabbit.

Minerals

Minerals are also important nutrients for optimum body growth, production and reproduction in rabbits. They can be divided into two categories: (*a*) macro-minerals which are required in much larger quantities *e.g.,* calcium, phosphorus, sodium, potassium, magnesium and chlorine, and (*b*) micro-minerals or trace minerals which are required in much smaller quantities *e.g.,* manganese, zinc, iron, cobalt, iodine, copper, molybdenum, selenium and chromium. Although the above mentioned minerals are required by the rabbit, those most likely to be deficient are calcium, phosphorus, sodium and chlorine which later are found together as salt. In growing, pregnant and suckling rabbits the requirement is higher by 5% in the diet. Rabbit is very efficient to absorb calcium and phosphorus from their diet.

The dietary requirements of different minerals are discussed below.

Calcium

Calcium, the most abundant mineral in the body, combines with phosphorus forms the dense, hard material of bone and teeth. The uptake depends upon pH, phosphates, phytates, oxalates and fats present in the intestine lumen. As rabbit teeth are continuously erupting at the rate of two millimetres per week, it requires constant supply of calcium. It helps in intracellular and extracellular fluid balance, muscle contraction, blood clotting, nerve cell activity, hormone regulation as well as maintenance and stability of cell membrane. Dietary calcium requirement is approximately 5.0–10.0 gm/kg of diet. Its deficiency results in poor mineralization of the bone and teeth. Bones become fragile due to osteoporosis. Excess amount is excreted as calcium carbonate sediment in urine. Metastatic calcification may occur with high calcium diets. Lactating mothers when fed high calcium diet do not generally accept their offspring to suckle leading to starvation of kits and sometimes, death of whole litter occurs. Feed intake also decreases when pellets contain high calcium. This is due to the selective feeding habit of rabbits.

Phosphorus

Phosphorus is closely associated with calcium and forms a major constituent of bone. Dietary phosphorus levels affect calcium absorption as calcium binds with phosphorus to form insoluble calcium phosphate in the gut. The absorption and excretion also depend on vitamin D. It plays vital role in energy metabolism. Growing parts of plants and many cereals are rich in phosphorus. In rabbit, low phosphorus diet causes hypophosphataemia, increase urinary excretion of calcium and shows the symptoms of 'pica' or depraved appetite where they even chew wood, bones, cloths or other foreign materials. In growing rabbits, it results rickets and in adults, osteomalacia. The ratio of calcium to phosphorus is very important for the utilization in body. In rabbit, a low ratio of 1:2 or 1:3 does not affect bone calcification or growth rate unless dietary phosphorus level is high. If it's dietary concentration increases to more than 1% and calcium to phosphorus ratio falls below 1:1, then the bone density decreases. The dietary requirement of phosphorus in rabbit is approximately 4–8 gm/kg.

Amongst other macro-minerals, the dietary requirements in rabbits are magnesium-3.0 g/kg, potassium-6.0 g/kg, sodium-2–2.5 g/kg and chloride-1.7–0.2 g/kg.

Trace minerals

Trace minerals are also essential as cofactors in enzymes which are involved in metabolic processes. The addition of these minerals in feed at higher level improves nutrient digestibility, feed conversion efficiency, growth rate, protein and energy utilization and dressing percentage in rabbits. Some of the trace minerals (like copper) should be provided to the diet in chelated form for better absorption and unwanted mineral interactions in the gut. In rabbit, the dietary level of different trace minerals is as follows.

Zinc	-	50 mg/kg
Iron	-	30–100 ppm
Copper	-	5–20 ppm
Manganese	-	8–15 ppm
Selenium	-	0.05 ppm
Iodine	-	0.4–2.0 ppm
Cobalt	-	1.0 ppm

BASIC PRINCIPLES OF RABBIT FEEDING

The following factors are to be considered during feeding in rabbits-

- Include fresh grasses and legumes to a level of about 70%. Provide 50% legume and 50% grasses as a minimum dietary requirement for resting does and bucks. Add a small quantity of hay to provide bulk and to prevent boredom of similar type of ration.
- Supply *ad libitum* concentrate to young or growing rabbits at all the times. Bucks, dry does and replacement stock should be fed at least once a day @100–200 gm concentrate feed.
- Always supply balanced ration to avoid nutritional deficiency.
- Supply clean and fresh drinking water throughout the period.
- Avoid abrupt change in quality and quantity of feed because the rabbit relies on a stable population of "good" bacteria in its gut and any change in dietary intake should be introduced gradually to avoid sudden change in types of bacteria present, as this can result in diarrhoea.
- Avoid dusty feed which can cause respiratory problem leading to even death of animals. Offer pellet concentrate feed preferably 3–4 mm in diameter and 10–15 mm in length. Only during emergency situation mash feed may be given after reducing dustiness through mixing water. Molasses @ 5% also may be used to reduce dustiness of feed as well as to improve the palatability of feed.
- Detoxify the feed ingredients before using in the concentrate ration. Do not provide sour or decomposed feed.

- Fix the feeding time in a day to avoid unnecessary stress to the animals.
- Do not provide heavy diet to the does just after kindling. Amount of diet should be increased gradually so that they can get full diet after about one week.
- Do not provide extra calcium in the diet of doe which interferes the acceptance of does to their kits and ultimately, affects the growth of litters.

Feeding Methods

Feeding methods of house rabbits are very simple and feeding is normally based on *ad libitum* fodder/grasses readily available nearby with little concentrate supplements. Roots, tubers and fruits are also offered without much process. In organized government farms or large scale commercial rabbitries, feeding methods are quite different. The following two methods are generally practiced.

1. ***'Limited' feeding*:** In this method, daily a measured amount, rather less than normal intake, of forages and pellets are offered in a container every day. Depending upon the consumption the amount may be changed in the following days. The advantages of this feeding methods are: (*a*) less wastage of feed, (*b*) less contamination of feed, (*c*) more consumption of feed due to freshness of feed, and (*d*) opportunity is there to check whether the rabbit has taken feed or not.
2. ***'Free choice' feeding*:** Unlimited feed is available to the rabbit at all times. Rabbits take *ad libitum* in a day. Several days pellets and hays are put at a time into the hopper or self feeder. After few days total feed intake is noted and intake per day is calculated. The advantages are: (*a*) saves both time and labour requirement, and (*b*) feed requirement is less. If green fodder feeding is practiced, it should be offered daily after cleaning the rack.

The feed container must be cleaned daily before putting feed into it. Normally, concentrate is offered one time in a day preferably in the morning at around 9 AM and green fodders first time at 11 AM and second time at around 4 PM.

FEEDING OF DIFFERENT CATEGORIES OF RABBIT

Rabbits are mono-gastric herbivorous animals. They require varieties of protein, carbohydrate, fat, fibre, vitamins, minerals, and water in their diet. Care must be undertaken in regular feeding to maintain proper health. Feeding rabbits a '*one size fits all*' style of rabbit food compromises their performance as the rabbits of different stages of life have different nutrient requirements. Four '*critical*' periods of rabbit feeding are (*a*) one week before breeding/mating, (*b*) one week after breeding/mating, (*c*) one week before kindling, and (*d*) three weeks after kindling. However, feeding of different categories of rabbits is described (Table 9).

FEEDING OF KITS UP-TO WEANING

Feeding of kits during the first two weeks of their life is simple. It is mainly imposed by the dam. Their eyes remain close up-to 9 days and ear up-to 11 days of age. During this time pheromone guides the kits to suckle their mother. Moreover, rabbits are known as excellent mother for nursing their babies. After giving birth rabbit comes to feed the kits in the nest box. Rabbits feed their offspring only once, rarely twice in every 24 hours, and that too, just for 5–10 minutes preferably before dawn or just after dusk when the doe feels safe. Out of which actual suckling phase lasts only for 2–3 minutes. In this short time frame, the kits get their whole day's milk supply. Round and distended belly is the indicator of proper suckling of kits. If the babies are off fed, they show sunken belly, wrinkled skin arising from dehydration and they become weak. Disturbance from the surrounding affects the nursing of young ones. Therefore, noise must be restricted in the area.

The fulfilment of nutritional requirements from birth to three weeks is reflected by the composition of their mother's milk including colostrums. Immediately after complete kindling, usually requires half an hour, the kits are allowed to suckle their mother to receive colostrums. The first milk of doe after, parturition, known as colostrum, has high protein, vitamins and minerals. The gamma-globulin of colostrum helps kits to build up body immunity to protect from many infectious diseases. It also helps to expel meconium, the first faeces after birth, as it has laxative effect. However, the milk quality depends on diet composition of lactating doe, amount of milk produced, amount of feed/ fodder consumed, quality of ration etc. The fat content of rabbit colostrum milk is greater than 17%. By the fourth day, it drops to around 10% which further steadily increases to over 15% by the eighteenth day. After third week of lactation, fat content in milk begins to taper off to less than 13%. But, the lactose content of rabbit milk varies from 1.1–1.7%. Thus, mother with high fat diet produces more milk resulting faster growth of kits. The peak milk production of does occurs around third weeks of kindling and after that, production reduces gradually. To maintain the consistency in milk production up-to weaning of kits proper feeding of mother is very important. The average composition of rabbit's milk is presented in Table 10.

After two weeks, put some amount of green fodder and pellet in a corner of the box where the babies can easily reach to acquaint with the feed and nibbling fodders. By the third weeks baby rabbit starts to consume appreciable amount of solid food. The feeding of baby rabbit is quite

Table 10: Average composition of rabbit's milk

Components	*Rabbit's milk (days 4–21) Percentage*
Dry matter	26.10–26.40
Proteins	13.20–13.70
Fats	9.20–9.70
Minerals	2.40–2.50
Lactose	0.86–0.87

Source: Lebas (1971)

different from adults. To promote faster body growth of baby rabbits 'creep' feeding is a general practice. Creep feed contains very high protein (22% CP), fibre (18% CF) and fat (5% CF). This feed is only accessible to the kits in the wooden part of nest box, not to the lactating doe generally kept in the wire part of nest box. Kits use to consume 40–50 gm of pellets at the age of 42 days. Sufficient water must be available for all the times. Young kits cannot sustain with sudden change in diet. Gradual replacement of feed ingredients is suggested to avoid any digestive disorders. The weight of kit at six weeks becomes around 800–1200 gm.

If doe dies during kindling it is very difficult to rear the kits for first two weeks. Use of foster mother is an alternate to save kits. Do not allow more than 3 kits per foster mother. The maximum age difference between the foster doe's litter and the fostered young should be 48 hours. Fostering should be done within 5 days of kindling. Otherwise, doe will not accept the young. If fostering is not possible, feed artificial colostrums individually after mixing of whipped egg- half, castor oil-half teaspoonful, goat/sheep milk-200 ml, honey-10 drops and distilled water-100 ml. To safeguard against aspiration hold the kit in an upright position or at a 45 degree angle and point the syringe of artificial colostrums down towards bottom or side of the mouth. For the first few feedings place the tips of the syringe in the corner of the mouth placing just a drop or two inside the mouth and wait to swallow. Use extreme caution not to get colostrum in their nose. If this happens, firmly press down on the bridge of nose of kits, and move down to the top lip and wipe the milk away with a soft tissue paper. No more than 4–5 ml per day per 6–12 hours in first week, 10–14 ml per 6–12 hours in second week, 14–26 ml per 8–12 hours in third to fourth week, and 26–30 ml per 12 hours in fifth to sixth week should be fed to the young ones.

FEEDING OF YOUNG/GROWING RABBITS

Feeding of growing rabbit depends upon whether they are reared individually or in groups. Growing rabbits, raised in a group, are always fed *ad libitum* or free choice. In groups they use to consume more feed and fodder than individually fed. It may be due to the competition in feeding between them. However, after weaning rabbits should generally be provided with more green fodders, hay and concentrate. On an average rabbits of 12 weeks of age can consume 80–100 gm pellet and 150–200 gm green fodder which increases gradually towards their adulthood. The intake of one adult rabbit is about 120 gm pellet and 250–300 gm green fodder.

FEEDING OF PREGNANT AND LACTATING RABBITS

Pregnant rabbits and lactating does both require high energy and other essential nutrients. They perform better when the energy comes mainly from carbohydrates instead of fat. The 'flushing' or feeding a higher energy feed just prior to breeding of doe also improves fertility and conception rates. The pregnant does should be supplied good quality green fodder and pellets she can eat. The bulk of the energy is needed particularly in the last one to two weeks of pregnancy due to rapid foetal growth. Reduce the amount of feed given to a doe 24 to 48 hours before she kindles to prevent caked mammary glands. After kindling gradually increase the amount of feed such a manner that the doe is full fed in seventh days. Normally one pregnant or lactating rabbit consumes about 150 gm pellet and 150–250 gm green fodder daily.

FEEDING DRY DOES AND BREEDING BUCKS

Dry does, breeding bucks and developing junior does do not require high energy diet that growing, pregnant and nursing rabbits require. Breeding rabbits should be reared '*not- too-fat-not-too-lean*' body condition. A little amount of good quality concentrate and a lower carbohydrate with high fibre can fulfill the requirements for sustaining growth of breeding bucks and mature does. It is very important to prevent the rabbits from becoming fatty. Many breeding failures are caused by does being overweight. In general, it is best to have a doe a little lean rather than a little fat. For junior does, the feed intake must be managed so that they will grow and be in good condition at the time of breeding. The intake of does should be restricted just after breeding. They are full fed only after confirmation of their pregnancy. If the doe fails to conceive, rebred and maintain her under restricted diet until becomes pregnant.

FORMULATION OF PELLET FEED FOR RABBITS

Rabbits of different categories must be supplied 'balance' concentrate pellet to fulfill any nutrient deficiency. Depending upon the availability of the fodders or grasses or hay the concentrate pellet is to be prepared. The required CP or DCP or TDN or DE can be adjusted at the time of formulation of pellet. The Pearson Square or box method is a very simple procedure that is being used for many years. Several numbers are put in and around the square. The most important number is put in the middle of the square which represents the nutrient requirement of an animal for a specific nutrient like CP, DCP, TDN, DE etc. Since protein is the most expensive component of animal feed, it is usually considered first followed by TDN or DE while computing rabbit pellet feed. The detail of procedure is described below.

Important Considerations for the Calculation

- The Pearson square ration formulation procedure is designed for simple rations.
- Nutrient contents of ingredients and nutrient requirements must be expressed on the same basis *i.e.,* 'dry-matter' or 'as-fed'.
- The value in the middle of the square must be intermediate between the two values that are used on the left side of the square.
- Ascertain the nutritive value of the feed ingredients to be used.
- Disregard any negative numbers that are generated on the right side of the square. Be concerned only with the numerical differences between the nutrient requirement and the ingredient nutrient values.
- Subtract the nutrient value from the nutritional requirement on the diagonal and arrive at a numerical value entitled parts. By summing those parts and dividing by the total, you can determine the percent of the ration that each ingredient should represent in order to provide a specific nutrient level. Always subtract on the diagonal within the square in order to determine parts. Always double check calculations to make sure

that you did not have a mathematical error. It also is very important to work on a uniform basis. Use a 100-percent dry-matter basis for nutrient composition of ingredients and requirements and then convert to an as-fed basis after the formulation is calculated.

Methods

1. Ascertain the CP content of the feed ingredients to be used in ration.
2. Separate the feed ingredients into two group I or protein source, and group II or energy source or *vice-versa.* Grouping is done by experience. Take the average value of CP for each group.
3. Calculate the ingredient to prepare 100 kg feed.
4. Calculate the total of compulsory components to be added in the pellet *e.g.,* mineral mixture-2%, salt-1% and vitamin powder (AB_2D_3)-0.2%. Subtract the total amount (3.2%) from 100%. Calculate the exact CP% to be present in the remaining amount (100 – 3.2 = 96.8%) of feed so that when these compulsory components are added, final mixture will have the required CP%. Sometimes, molasses (5%) is also considered as essential component in rabbit pellet because of its good binding ability of pellet.
5. Draw a square and write the calculated CP% in the middle of the square.
6. Place group I with its CP% at the upper left hand corner and group II with its CP% at the lower left hand corner.
7. Ignoring any sign calculate the difference between CP% of group I written in the left upper and calculated CP% written in the middle of square. Write the difference on the diagonal of right lower of square. Similarly, write the difference of group II and calculated CP% on right upper of square.

8. The amount of feed under different groups is then expressed in percentage basis. Thus, group I% = (X × 100)/(X + Y) and group II% = (Y × 100)/(X + Y)

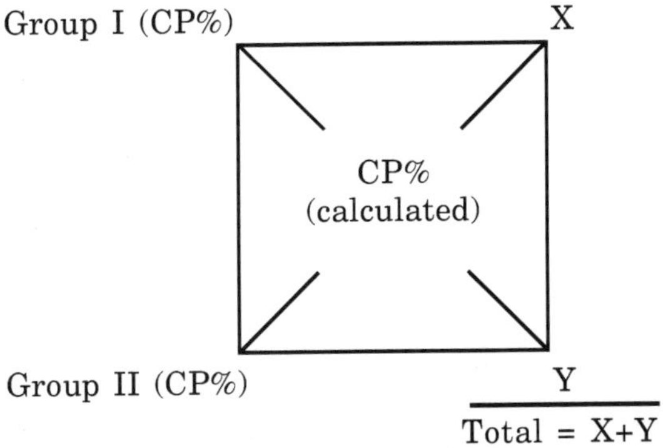

9. Calculate the amount of different ingredients from each group dividing total amount of the particular group to the number of ingredients to be used in that particular group. Experience persons use the ingredients from different groups considering the market cost and availability of feed to formulate low cost ration. It is always suggested to use more number of ingredients from different protein and sources. It improves the palatability of feed as well as fulfills the deficiency of each other ingredient.

Example 1: Prepare 100 kg rabbit pellet feed with 16% CP using crushed maize, GNC, crushed barley, meat meal, rice polish, mineral mixture, salt and vitamins through Pearson-Square method.

Solution: The CP% of the ingredients to be used in pellet are Crushed maize-10%, GNC-42%, Crushed barley-11%, Meat meal-50% and Rice polish-12%.

The essential components of rabbit pellet are 2% mineral mixture, 1% salt and 0.2% vitamins. To prepare 100 kg pellet with 16% CP, the calculated CP will be = 16 × 100 / (100 – 3.2) = 1600/96.8 = 16.53%

Average CP% of group I (energy source)

= (10 + 11 + 12)/3 = 11%

Average CP% of group II (protein source)

= (42 + 50)/2 = 46%

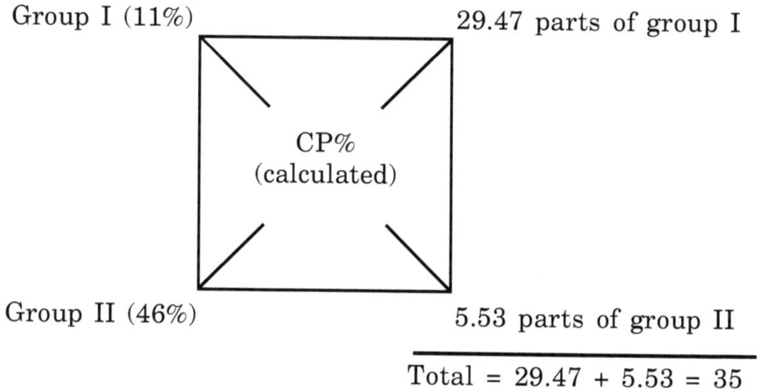

Or, depending upon the market cost of feed it may also be -

Crushed maize	=	25 kg
Crushed barley	=	22 kg
Rice polish	=	34.5 kg

Similarly, the individual ingredient to be used in pellet from group II

GNC	=	15.3/2 = 7.65 kg
Meat meal	=	7.65 kg
Or, GNC	=	10 kg
Meat meal	=	5.3 kg

Due to high cost and off flavour the protein level of each ingredient from animal source should generally be restricted up-to 10%. It may increase or decrease according to different situation.

Therefore, final composition of 100 kg rabbit pellet will be as follows:

Crushed maize	=	25 kg
Crushed barley	=	22 kg
Rice polish	=	34.5 kg
GNC	=	10 kg
Meat meal	=	5.3 kg
Mineral mixture	=	2 kg
Salt	=	1 kg
Vitamins	=	0.2 kg

Chapter 4

Reproduction and Breeding Management

Based upon the fact of excellent prolificacy of rabbits, there is a proverb *'to breed like rabbits'*. Their number can be increased at very faster rate due to short gestation period, short service period and short generation interval, large litter size and not having any specific breeding seasons. The reproductive cycle of rabbit is very peculiar. The proper knowledge of reproduction is necessary to get the optimum utilization of this species. The does do not come into oestrus like other mammals. The interesting thing is that without buck the doe can be ovulated even through the mechanical stimulation at vagina even with the help of simple glass rod. Hence, the rabbit is known as *'induce ovulator'*. It is possible for a doe to accept a buck for breeding at any time of the year. The rabbits are very much protective of their own domain. For successful mating doe should always be taken to buck only as the doe will sit in buck's territory being feeling her in an insecure domain. However, before going into the rabbit reproduction, it is very essential to acquire basic knowledge about the anatomy of male and female rabbit reproductive organs.

MALE GONAD DEVELOPMENT, PUBERTY AND SPERM PRODUCTION

In male, gonad differentiation starts from the 16th day of fertilization. The development of testes from birth to five weeks is comparatively slower than other body parts. After that, testes and accessory glands grow very rapidly. In between 40 to 50 days of birth spermatogenesis starts while testicular tubes become active at 12 weeks of age. At about 110 days first spermatozoa are present in the ejaculate. The sexual behaviour is observed even at 60 days of age. Although the occurrence of first coitus is found at the age of 100 days, but the viability of sperm cell of first ejaculate is either very poor or near nil. The male should generally be allowed to breed for the first time at the age of 5 months. However, the onset of puberty may be attained little earlier or later depending upon the breed, nutrition level, climatic condition, husbandry practices etc.

The buck normally ejaculates 0.3–0.6 ml semen with 150–500 × 10^6 spermatozoa per millilitre. The sperm concentration varies from age, false mounting, successive matings, frequency of use of buck, nutrition, interest in does etc. In case of two successive services, first ejaculation contains more volume and low concentration while second bears reverse attributes. The concentration decreases after two successive matings. False service results high sperm concentration. Buck produces around 150–300 × 10^6 spermatozoa per day. Considering this, buck should be used once in a day for maximum production of spermatozoa. Daily two times regular use lower the spermatozoa concentration into half for each case. About one million live or two millions total spermatozoa (minimum 50% motility) in 0.5–1.0 ml diluted semen are required for conception.

FEMALE GONAD DEVELOPMENT, PUBERTY AND OESTRUS CYCLE

After sexual differentiation at 16th day of fertilization, ovogonial division starts on the 21st day of foetal life which continues up-to birth. The first follicles appear at 13th day of age and the antrum follicles at about 60–70 days. The doe shows the sexual behaviour at quite early age even around 70 days, but that should not be considered as age of puberty. If the does are bred also, they do not conceive because at this stage ovulation does not occur. The female rabbits are bred for the first time when they attain at least 70% of their mature body weight. The onset of puberty in female also depends upon many factors as described in male.

Like other mammals female rabbits do not have any specific oestrus cycle with regular heat period. Females are considered to be in oestrous when she accepts service and in dioestrus when she refuses to mate. Few researchers are in the opinion that the oestrous cycle of rabbit ranges from 16–18 days. Out of which 12–14 days the does accept the males and remaining 4 days do not. Practically, it is found that the doe with red swollen moist vulva generally accepts service. Ovulation occurs 10–12 hours after mating. The doe can stay in heat for several days running. On the ovary, follicles not having evolved to the ovulation stage through lack of stimulation undergo regression and are replaced by new follicles which remain for a few days in the pre-ovulating state and may then in turn regress.

SEXUAL BEHAVIOUR AND MATING

The wild male and female rabbits generally mark their territory to establish their social hierarchy and to show their mate that they are ready to copulate. The chinning behaviour in which the animals rub their chins against an inanimate object while giving off odoriferous substances with their submandibular or chin gland. This behaviour is also found in domestic rabbits when reared male and female together in groups. If female is taken to male cage for mating for a period of time chinning behaviour is not prominent. However, normally the male rabbit exhibits sexual behaviour like chasing of female, tail flagging in which the buck parades before the doe with tail elevated over the back, attempts to copulate, and amatory behaviours *i.e.,* grooming and licking. In a display of sexual aggressiveness the buck projects a jet of urine at the doe. It is also interesting that when young buck is allowed to mate to adult doe the female chases the male vigorously leading to '*jousting*' or formalised leaping which generally occurs between two males. On the other hand, the female counterpart shows red swollen moist vulva along with the typical pose to accept male called '*Lordosis*'. The doe raises her hindquarters to allow copulation. The buck mounts on the doe by grasping the body of doe with forelegs on the flanks. After a number of quick copulatory movements with pelvic thrusts ejaculation follows on the first intromission of penis into female genital tract. The copulation cry is a common in rabbits which may arise from either of the partners. It can be considered as an indication of successful mating. At the time of ejaculation both the rear feet remain off the ground. As a result, the buck losses his balance and falls either left or right or back side of the doe. The sex drive of rabbits depends on heredity, breed, ambient temperature, age at breeding, mating frequency, nutrition, body condition etc. In an organised rabbitry, allow one buck to mate 2–5 does in a week for optimum performance. To avoid dire consequences of inbreeding in the farm more number of unrelated bucks should be used. For this, keep one buck for every 10 does or even less.

Fertilization

About an hour and a half after ovulation, the ovocytes become fertilizable. On the other side, the sperms are deposited by the male during coitus in the upper part of vagina which move rapidly towards the upper side to reach fertilization area *i.e.,* distal ampula within 30 minutes of coitus. The maturation process is going on the route itself.

Out of 150–200 million spermatozoa, only 1% reaches in the uterus. Rest are destroyed in the uterotubal junction and cervix. After 72 hours of ovulation egg reaches the uterus. Implantation takes place seven days after coitus, at the blastocyst stage. These blastocysts are fixed at almost equal distance in each uterine horn.

PSEUDOPREGNANCY

When the ova are released but not fertilized, the pseudopregnancy condition occurs which lasts upto 15-18 days of ovulation. In this case, the corpus luteum and uterus develop as in normal pregnancy but do not reach their size or the level of progesterone production of the corpus luteum in pregnancy. Even they prepare the nest with fur but never give birth. Due to release of prostaglandin from uterus near 12th day of ovulation the corpus luteum regresses and disappears from the ovarian surface. This exceptional situation occurs when the doe is mated with a sterile but sexually active buck or doe is serviced under unfavourable conditions or is inseminated by poor quality semen or is serviced with very young buck.

PREGNANCY

It is well established that the progesterone hormone is mainly responsible for the maintaining of pregnancy in animals. This hormone secretion continues to increase from third to fifteenth day of coitus, then remains stationary and finally, drops quickly before kindling. The normal gestation period in rabbit is 30 ± 2 days. It varies from breed, sex of foetus, litter size, litter weight, season and nutrition.

PREGNANCY DIAGNOSIS

Diagnosis of pregnant doe is very important for successful

rabbit farming. The objectives are to confirm the pregnancy of doe, to manage the doe according to their pregnancy, to reduce wasteful expenditure through rebreeding, to know the expected date of kindling, to maintain proper breeding cycle in the farm for maximum crop in a period of time, and to plan sell or purchase or slaughter of animals. A number of methods are followed to determine the pregnancy in rabbits which are described as under.

1. After restraining the rabbit with one hand, place another hand between the hind legs in front of the pelvic region just a little below the abdomen. The viable embryos will be palpated by the fingers and thumbs at 10–12 days after mating.
2. Pregnancy can be diagnosed by observing the symptoms like swelling of udder and teat, and vaginal seal. It occurs due to the hormonal change in the pregnant doe. It is easily diagnosed by the experienced persons.
3. Take the doe to the buck after 4–5 days of mating. Normally, the buck will not mate the pregnant doe. Sometimes, the aggressive buck forces to mate the doe which mated 4–5 days back. If the doe does not allow mating easily, the doe is likely to pregnant.
4. The pregnancy of doe can be determined by gradual increase in body weight change at different weeks after mating. Diagnosis is more accurate in older doe compared to first time pregnant rabbits. Because, the weight changes of growing rabbits is confused with the weight changes due to pregnancy. If the young rabbit becomes pregnant, the increase of body weight change is more than the adult pregnant rabbits.
5. The pregnancy can be confirmed at 7 days after mating through ultrasonography. Due to high cost involvement of ultrasonoraphy equipments, this is not feasible for common farmers.

POST-PARTUM BREEDING

In rabbits, uterine retracts very quickly. It is interesting to note that the uterus looses more than half of its weight within even less than 48 hours of kindling. The doe can be rebred three days after kindling. But, it is not recommended. Some of the stockpersons also bred them two weeks before weaning. However, to achieve better conception rate and to get healthy kits at weaning rebreeding of doe should be practiced after weaning of kits *i.e.,* 42 days of kindling. The breeding cycle of rabbit is presented in Fig. 17.

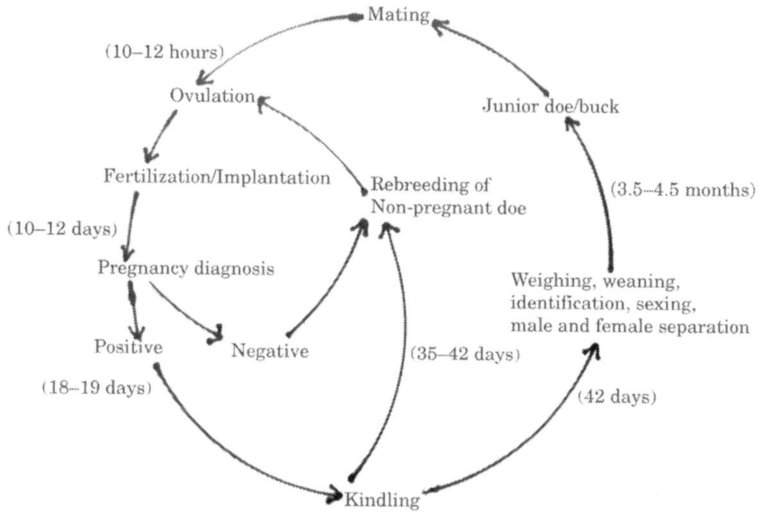

Fig. 17: Breeding cycle of rabbit

KINDLING

The act of parturition in rabbits, the kindling, is a very complicated phenomenon. It is believed that the secretion of corticosteroids by the supra-renal of kit gives signal to initiate parturition. The prostaglandin $F_{2\alpha}$ also helps this process. Towards the end of gestation the doe makes nest for the babies with her own fur or other available bedding materials like coconut fibres, straws, dry grasses, wooden shavings, paper cutting etc. Depending upon the litter size kindling completes within 10–30 minutes. Litter size normally varies from 3–12 or even more. Depending upon breeds, litter size, age of doe, nutrition etc. the birth weight of kit ranges from 40–60 gm. The higher birth weight of 120 gm or even more is also recorded especially when litter size is very low. The doe gives birth any time but preferably at night.

MANAGEMENT OF DOE AROUND KINDLING

The kindling box should be cleaned and disinfected before putting bedding materials within it. The depth of bedding materials depends on the environmental temperature. Then, shift the pregnant doe 4–5 days before the expected date of kindling. Feed and fodders should be available for all the times. The shed must be very calm and quiet around kindling. Unnecessary chasing and rough handling are to be avoided to reduce stress to doe. Closely observe the doe for the appearance of symptoms of kindling and wait for complete parturition. In case of dystocia, oxytocin injection may be helpful. Immediately after parturition, clean the placenta from the cage to avoid intake by the doe which may cause drastic reduction of milk production and ultimately, leads to death of kits. After cleaning vulva and adjacent area with mild antiseptic solution allow the doe for about 5–10 minutes to suckle her babies. The doe should be separated after milk feeding. Little amount of concentrate with luke warm water is to be provided to the mother. Full feeding of mother immediately after parturition is generally avoided. Rather, amount of feed should be increased gradually in such a way that on seventh day of parturition the doe will be full fed.

MANAGEMENT OF KITS AFTER BIRTH

After complete kindling separate the doe from the nest box. Normally, the kits are born with their eye and ear closed and nearly hairless. At this stage proper care is must to protect these kits from adverse climatic condition. The kits are to be placed in the right place in the nest box with appropriate bedding materials. If not, they should be confined together with their litter mates. Count the babies after birth, eliminate the dead and deformed kits. Special care should be taken for runts. Take the weight of each kit as well as whole litter. To remove the human smell from the bare hands of handler rub the hands with the nest bedding materials before handling kits. Allow the doe to lick her babies to remove mucous membrane present in the nose including body of kits, and to improve peripheral blood circulation of young ones. Complete these jobs at the earliest and allow the doe in the nest box to feed her babies. If the bellies of kits are rounded, it is confirmed that they have suckled their mother. The mother is generally separated from the nest box after complete suckling. In case of death of mother at parturition, either foster mother or artificial colostrums is used to save the kits. After 10th day of birth, if eyes are not open, take a cotton swab with distilled water and clean the area of eyes to open. The nest box must be protected from the rat, mice or other insects. Do not ever keep a deep water container in which a baby could drown. Instead, use something shallow which can be ringed and filled frequently.

IDENTIFICATION

The identification is a process in which the animals are marked for their individual identity. It may be either temporarily or permanently. The tagging and tattooing are the mostly preferred identification system in rabbits. Whatever the system is followed, it must be as permanent as possible, visible and readable from a particular distance, economic, not harmful to animals, and acceptable to the stockpersons. Identification is generally done in rabbit at the time of weaning *i.e.,* 42 days. It may be changed considering the weaning system adopted in the farm. The identification is helpful to confirm the ownership of animal, to do insurance of animals, to maintain different kind of records, to issue any type of certificate against them, to prepare post-mortem report, and to get bank loans.

Methods

Tagging

The tag is made up of either light metal or strong plastic and is of different shape. The number may or may not be stamped on it. The plastic tags are generally available at the market with numbers stamped on it. When plain plastic tags are used numbers are to be written by permanent markers on the top of tag. Over the period of time the written number may be rubbed or disappeared from the tag. So, proper care must be taken to maintain this kind of tag. The metal tag is prepared from copper and aluminium. The numbers are printed with the help of number printing chisels. It is permanent in nature. These tags are fitted in the ear of rabbit with number placed upward side of tags. Special ear tagging forcep or machine is used for this. The numbers are tagged in either side of the ear. There is no hard and fast rule for this. Some stock persons use left ear for identifying female and right for male. Again, the tags are of two types *i.e.,* self piercing and non-self-piercing.

1. ***Self-piercing tagging*:** This type of tag has sharp end and can be directly applied to the ear through tagging forcep. In this method, the tag to be used and forcep are sterilised. The area of ear is cleaned by using spirit. The area of ear where tag is to be fitted must have

fewer veinules to avoid unnecessary bleeding. After securing the animal tags are fitted in such a way that the tag becomes neither too loose nor too tight in the ear. Appropriate gap between curved end of tag and edge of ear should be there considering the future growth of ear. Apply tincture iodine or antiseptic ointment along with fly repellent on the wound for quick healing of wound. Follow dressing until wound heals properly.

2. ***Non-self-piercing tagging*:** This method is almost similar as in case of self-piercing tagging. Here, instead of fitting the tag directly on ear, first one whole is created on the cleaned ear with the help of sterilized puncher and then, the tag is fitted.

Tattooing

The tattooing set, consists of tattooing forcep, metal pins in the shape of numbers and letters and tattooing ink, is available in the market which is used for this identification. First, the tattooing forcep, numbers and letters are sterilized and inner surface of ear of rabbit is cleaned with spirit. The desired numbers and letters are then pierced in the inner less veinule area of ear of rabbit through tattooing forcep. The tattooing ink is rubbed into the pierced area. The ink penetrates skin and gives impression on ear. Dress the wound properly for healing. This method can solve the purpose temporarily for 6 months as the number becomes hazy with the increase of age.

SEXING

The sexing is the identification of male and female rabbit. It is not a very easy job to differentiate between the male and female kits at early stage. Only experienced persons can do this. Under organized farm, sexing is done at the time of weaning preferably at 42 days of age. It can be earlier also. The purposes of sexing are to count male and female numbers in a litter, to manage separately for avoiding fights and to keep male and female individually for giving better management. The process follows the holding of rabbit in the palm of the hand turn it upside down with the tail between fingers. Apply gentle pressure on each side of the vent which is the area including the genitalia *i.e.,* opening of reproductive tract and the anus *i.e.,* opening of digestive tract. The anus will be the closest and genitalia, the farthest from the tail. In male, penis will appear as a tubular protrusion with a small round opening at its centre. It is round in diameter and light pink in colour. Male rabbits of more than 10 weeks show pinkish folds of skin on each side rather slightly cranial to the penis. Female rabbit will have slanted oval protrusion on gentle pressure on vent area and have slit like structure at centre (Fig. 18).

WEANING

The weaning is the process of separation of mother from her babies. Like other animals mother rabbit is generally taken away from the nest box, not the young ones. If babies are shifted from nest box, babies undergo severe stress due to weaning as well as environmental stress. But, the doe faces only weaning stress. Weaning of rabbits is generally practiced at the age of 42 days. It can be done earlier also. The body weight of individual kit at weaning comes about 800 to 1200 gm. This is a very critical stage when 'weaning syndrome' can develop in kits because, the young ones are switched over to pellet and fodders, and are no longer depending on the mother's milk for their nutrition. Some kits can develop diarrhoea as they do not absorb the nutrients from the food the way they need to in order to stay healthy. It is to be confirmed that before weaning off the mother the kits are eating and drinking on their own.

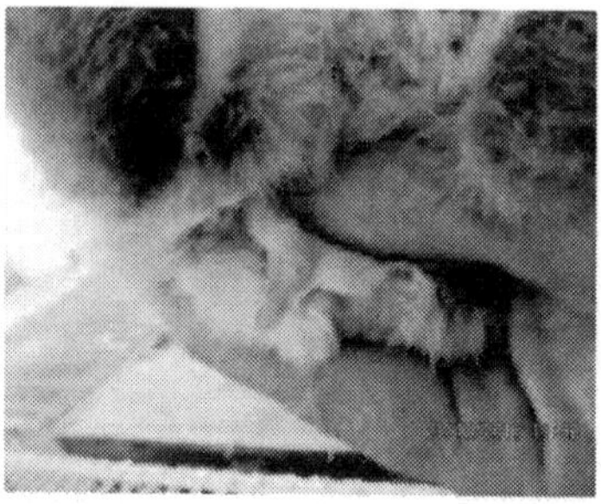
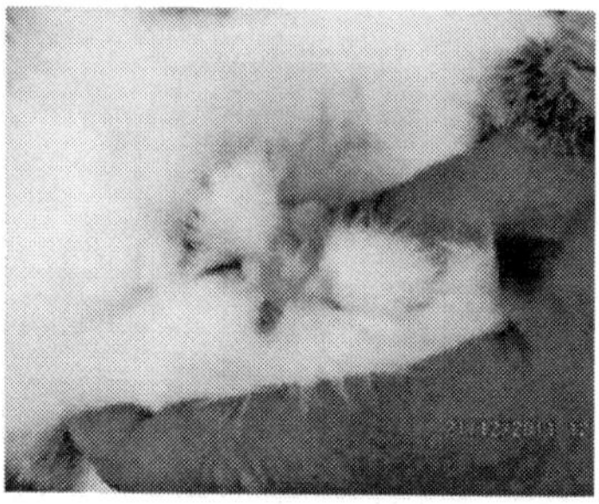

Young male

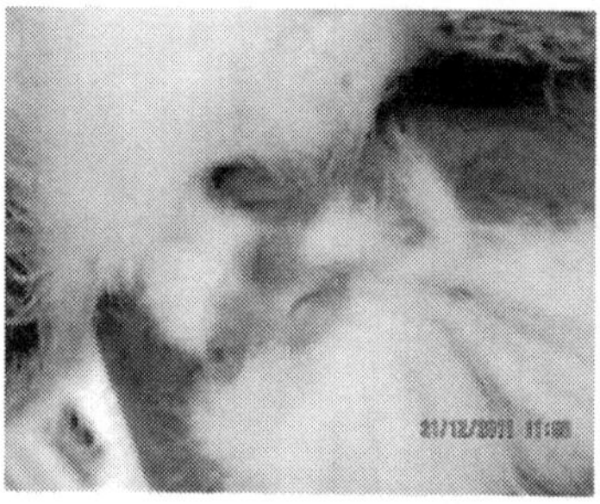

Young female

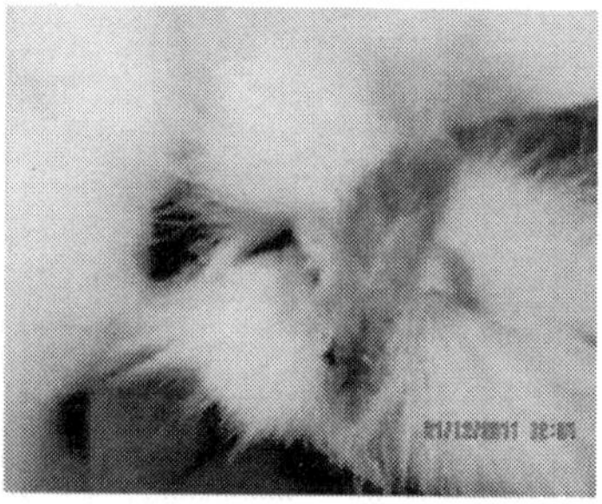

Adult male

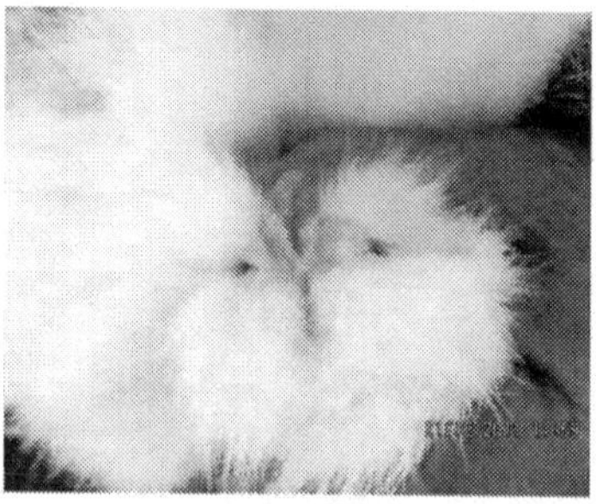

Adult female

Fig. 18: Sexing in rabbits

Weaning is very necessary to prevent the fighting of male and female rabbits keeping in a single cage.

SELECTION OF BREEDING RABBIT

The process of choice of individual to be used as a parent and will produce offspring in the next generation is called selection. Selection plays crucial role in improving the performance of animals. It is a continuous process starting from birth to different stages until maturity of individuals and thus, is considered a multistage process. Breeding of rabbit through appropriate selection can result higher genetic gain within a short period. But, faulty selection deteriorates the performance of animals very quickly. The selection is done to identify the superior germplasm in the base population, to cull or sale or purchase of animals, to obtain higher genetic gain, and to improve low producing animals. The different economic traits in rabbits are:

1. Litter size and weight at birth
2. Litter size and weight at weaning
3. Body weight at marketing *i.e.,* at 12 weeks
4. Feed conversion efficiency
5. Mothering ability
6. Disease resistance
7. Reproductive efficiency
8. Wool quality

Simply, the selection is nothing but the comparison between animals. It must be within the same species, breed, sex, age group or stage of life. Moreover, based upon the available parameters the selection criteria may be changed to some extent according to the place *viz.,* at farm, owner's house, and market or animal fairs. The animals are generally selected on the basis of individual records, pedigree records and on other economic traits. The rabbit must be true to the selected breed. Like females, males are also to be given equal importance during selection. The male is considered as "half of the herd" as he transmits half of the genetic materials to his offspring in the next generation.

The breeding rabbits are selected considering the following points.

1. ***General considerations:***
   - A. **General appearance:** Active, bright eye and healthy.
   - B. **Genetic and phenotypic abnormalities:** Absent
   - C. **Body conformation:** It depends upon the breed.
     - (*i*) *Head*: Well developed, larger ear to dissipate body heat during summer.
     - (*ii*) *Teeth and jaws*: Sound
     - (*iii*) *Back*: Long and wide
     - (*iv*) *Hip*: Wide, round and long
     - (*v*) *Legs*: Straight
     - (*vi*) *Hind feet*: Large
     - (*vii*) *Belly*: Compact and not pot in shape
   - D. **Growth rate:** Very fast
   - E. **Feed conversion ratio:** Very good
   - F. **Disease resistance:** High
   - G. **Carcass quality:** Excellent
   - H. **Fur quality and production:** Excellent
2. ***Special considerations:***
   - A. **For buck:**
     - (*i*) *Testes, penis and prepuce*: Well developed without any adhesion of penis with prepuce.
     - (*ii*) *Sex libido*: High
     - (*iii*) *Semen quality*: Very good
     - (*iv*) *Semen quantity*: Sufficient
     - (*v*) *Fertility*: Record of producing large litter.
   - B. **For doe:**
     - (*i*) *Teat*: Eight evenly placed well developed teats.

(*ii*) *Litter size*: Large

(*iii*) *Litter weight*: More

(*iv*) *Mothering ability*: Very good (Making good nest, litter size and weight at weaning are the indirect indicators of good mothering ability of rabbits).

(*v*) *Breeding*: Ability to conceive at early age, high conception rate and easy rebreeding.

(*vi*) *Body condition*: No tendency to become fatty.

BREEDING SYSTEM

I. Outbreeding II. Inbreeding

I. Outbreeding

Mating between distantly related or unrelated (by ancestry for 3–5 generations) animals is called '*outbreeding*'. It increases heterozygosity. Different types of outbreeding are:

Crossbreeding

Mating of animals from different established breeds is called crossbreeding.

Advantages

- It introduces desirable character into a breed in which they have not existed formerly.
- It aids in the formation of new strains/breed.
- Crossbreed animals exhibit accelerated growth and vigour.
- It causes quickest as well as greatest improvement in most of the economic traits.
- It results quicker replacement of every low-yielding animal.

- It takes advantages of good qualities of two or more different breeds.
- It maximizes both heterosis and frequency of desirable genes with additive effect.

Disadvantages

- It is necessary to maintain two or more purebreds stock to raise crossbreeds.
- To achieve genetic gain accurately rigid selection pressure with crossbreeding strategy is to be given.
- It needs more space and capital investment.
- This system breaks up established characters and destroys combination of characters.
- By this method, total performance of animal in all aspects is not improved but for only a single or few traits can be improved.
- It is difficult to sell the new breeds to producers even if it is having real merit.
- It is difficult to attain uniformity of an established breed among progeny of a crossbreed foundation.
- Use of crossbreed animals for breeding purpose is having wide segregation of economically important traits and a general decline in vigour below that of parental breed.

Various types of crossbreeding are described as under.

1. ***Two breed crosses*:**
   - A. **Two pure breed crosses:** Here, two pure breeds are mated together and their crossbreed progenies are not used in the breeding system.
   - B. **Inter-se mating:** Crossing of cross breed progenies having the same level of inheritance of the two breeds *e.g.*, crossing of F_1 with F_1.

C. **Back crossing:** It is the mating of crossbred animals to pure breed animals of either breed used to produce crossbreed progeny of first generation (F_1) *e.g.*, F_1 x P_1 or F_1 x P_2 where P_1 and P_2 are two purebreds used to produce F_1 progeny.

D. **Crisscrossing:** It is similar to back crossing except that both parental breeds (P_1 and P_2) are used alternatively in each successive generation. When A and B two breeds are used in crossbreeding, crisscrossing with them will be like a following manner.

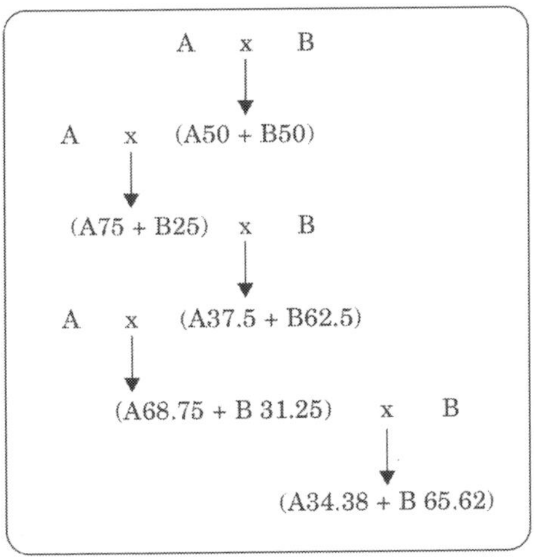

2. ***Three breed crosses (triple crossing)*:** In this system, three breeds are used. The males of three breeds are used in rotation in the successive generations of crossbred females of the previous generation. This is thus called '*rotational crossing*'.

3. ***Four breed crosses (double two way crossing)*:** Crossbred females produced (AB) by crossing two breeds (A and B) are mated with crossbreed males produced

(CD) from crossing another two breeds (C and D). Thus, the mating between AB and CD is called double two way crossing.

Crossbreeding is used/advised: (*i*) for low heritable traits like litter size and weight, increased vigour, reproductive efficiency etc., (*ii*) for quicker improvement of low yielding animals, (*iii*) for establishing new breed or strain, and (*iv*) for exploiting heterosis of traits of economic importance.

***Outcrossing*:** When mating is done among unrelated animals of the same breed, the breeding system is called '*outcrossing*' and the progeny is called 'outcross'. In this system, genetic variability exists within the breed. The outcrossing within a herd by use of selected sires is called as '*selective breeding*'.

1. ***Top crossing*:** It is the mating of inbred sire with non-inbred dam. This breeding is mainly used to exploit hybrid vigour by using reciprocal selection experiment. Principle of top crossing and grading up is same. Top crossing is usually applied to different families with in a pure breed, whereas, grading up is applied to continue use of sires of one pure breed starting with foundation dams which are of another breed or of no particular breed at all.
2. ***Grading up*:** In this crossbreeding, sire of a pure breed animal is mated to the dam of non-descript, scrub or native (desi) of another breed or of no particular breed at all. The crossbred female is back crossed to the pure breed sire. Thus, the produced offspring will have 75% blood from pure breed. After seven generations of crossing, the graded dam and purebred sire will produce the progeny which will have about 99% blood of pure breed.

Advantages

- After seven generations, pure breed animals can be obtained.
- It is less expensive and can be started with a little money in comparison to the purchase of an entire herd of pure breed animals.
- It helps to prove the genetic potentiality of sire.
- It is a good start who can slowly change over to pure breed animal.

Disadvantages

- Pure breed animals are always better than graded or desi animals for use.
- It is difficult to raise commercial stock as appropriate hybrid vigour or heterosis can't be noticed in upgrading experiment.
- Genetic improvement is more time consuming.
- Lack of adaptability is found in animals with high proportion of improved ancestry. For upgrading, it is advisable to use a breed that has thrived well under local conditions. Otherwise, graded animals may not adopt in the local environment.
- Number of desirable genes in pure breed animals for efficient production is very limited to use.
- Breeding value of individual animal may be reduced to a certain degree because they are heterozygous and do not breed as uniformly true as inbred animals.

3. ***Species hybridization:*** The crosses between two species of widest possible type of outbreeding is called '*species hybridization*'.

II. Inbreeding

Mating between closely related (by ancestry) animals is called '*inbreeding*'. Inbreeding has following consequences:

- It increases homozygosity and decreases heterozygosity and favours uniformity among animals. Thus, it fixes characters in an inbred population though increased homozygosity.
- It does not create any new gene and does not change gene frequency.
- It increases additive genetic variance.
- It decreases heritability in a small population.
- It does not increase the number of recessive allele in a population but brings them in light through increasing homozygosity.
- It results inbreeding depression leading to poor growth rate, reproductive performance, vigour and production traits, and high risk of abnormality.
- It results uniformity among animals within the line.
- It reduces intensity and effectiveness within line selection.

Methods of Breeding

1. ***Tandem method:*** After fulfilling the target of one trait, next trait is taken into consideration with target. In this method of mass selection the individual trait is improved successfully *i.e.,* selection is practiced for only one trait at a time till the satisfactory improvement in this character is achieved.

   Advantage

   - It is useful in the traits of high heritability.

   Disadvantages

   - It does not give satisfactory result when heritability of trait is low.
   - At a time more number of traits can't be selected due to negative correlation between traits.

- It takes longer time to improve all traits.
- Genetic progress per unit time is less.

2. ***Independent culling level methods*:** In this selection method, more than one character is considered. For each character lowest value and the target value are taken into account. Animal fails to meet this standard in any one of the trait is rejected.

   Advantages

   - At a time a number of traits can be considered.
   - It allows culling of animals earlier as it takes less time compared to tandem method to achieve targets.

   Disadvantages

   - Individual animal may be rejected due to highest value of one trait but lowest value of another trait.
   - When number of character is more than three, it is very difficult to estimate.
   - Early expressed traits get more weightage than late expressed traits.
   - It requires tedious work to determine optimum culling level.
   - Intensity of selection becomes low.

3. ***Selection index*:** Information of different economic traits of animals is collected from a family and an equation or index is constructed considering the importance of traits. On the other words, after allotting one score to each economic trait, an index of net merit of an animal is formulated for number of traits. This index value is considered as the actual value of individual animal. This selection can be applied to large number of characters at a time, hence, is called '*multi trait selection*'. This method is superior over tandem and independent culling level method when animals

are to be selected considering number of economic characters simultaneously.

BREEDING OF RABBIT

The rabbits have 44 chromosomes (22 pairs) in their cell. The genes are responsible for expression of various characteristics which are randomly passed to the progeny through breeding over the time. The breeding system followed in rabbits depends upon their utility. If they are used in experimental purpose, inbreeding *i.e.,* mating between closely related rabbits is done for producing uniform (homozygous) rabbits. It eliminates the individual variation between the animals to be used in different treatment groups. The out breeding *i.e.,* breeding between unrelated animals is followed when the rabbits are bred for any improvement. To maintain the purity of breed selective breeding is practised. When the aim is to get the benefits of numbers of desirable characters within the same breed crossbreeding is done. The body size, daily weight gain, feed efficiency, teat number, milk production, wool quality and quantity, coat colour have high heritability and crossbreeding can help to improve these characters. On the other hand, all reproductive traits have low heritability. Therefore, cross breeding is not beneficial for the improvement. In grading up, high producing bucks are crossed with the low producing does to enhance production and reproduction performance of animals.

Chapter 5

Housing Management

Housing and equipments are the essential components of rabbit farming. The basic requirements of rabbit shelter are to protect them from hot and chilly winds, to protect from strange animals and theft, to provide clean hygienic surroundings, to supply proper light and fresh air into the shed or within enclosures, and to prevent the rabbits from escaping. The house must facilitate self-cleansing, easy to maintain, durable and economic. Rabbits can tolerate cold up-to a certain extent but ambient temperature beyond 30°C is dangerous for their survival. The structure and design of the shelter depend upon the ambient temperature, humidity, rainfall, solar radiation etc. The house should be modified such a way to enable them calm and comfortable shelter which is necessary for achieving the optimum production and reproduction exploitation.

BASIC REQUIREMENTS OF RABBIT HOUSING

The rabbit house must fulfil following basic needs-

1. Calm and quiet.
2. Comfortable for rabbits throughout the year in different weather.
3. Protection from predators like cat, dogs, mice, rat, snake, insects, vermin, mosquitoes etc.
4. Escape proof.

5. Easy to clean, disinfect and fumigate.
6. Easy access to the workers
7. Economic to the stock persons and easy to maintain.

HOUSING ENVIRONMENT

Ambient Temperature

Rabbits maintain their constant body temperature. The most important fact is that the rabbits do not have functional sweat glands and breathing is an important control mean to dissipate latent heat from their body. To some extent ears of rabbit take part in regulating body heat. Its function is like a 'car radiator'. The ambient temperature affects feed intake, physical and physiological responses, and behaviour of rabbits. The rabbits use to change their feed intake according to the existing microclimate. Normally rabbits huddle together in low temperature and move apart in high temperature. Body reserve in terms of fat helps to maintain the environmental stress. Rabbits are very sensitive to hot weather comparative to cold. They can tolerate the atmospheric temperature between –20°C to 30°C. But, the ideal ambient temperature for rabbits to achieve optimum production and reproduction is usually 15°C to 20°C. Maintaining a constant and uniform temperature within the rabbit shed are the prerequisites of rabbit housing. Due to this reason, the compartment of rabbitry should neither be too big nor too small in size.

Relative Humidity

The rabbit is quite sensitive to the change of atmospheric relative humidity. Therefore, it should be maintained at constant rate within the shed. Ideally, the relative humidity within the rabbit shed should be 75% at 16°C ambient temperature.

Ventilation

The rabbit shed should be airy, free from any draughts, dust particles, and other harmful gases. Inlet ventilators are fitted for getting fresh air to the rabbits and outlet ventilators are fixed to evacuate vitiated air containing carbon-di-oxide, ammonia, hydrogen sulphide, methane etc. from the rabbit shed. Ventilators are fixed at the proper site depending upon the climatic conditions of the region. Hopper or centre pivoted window ventilators are most widely used in enclosed sheds. Under tropical climate, water coolers and ceiling fans can be used to regulate air circulation within the enclosure particularly in summer months. It may not require in temperate climate. Permanent shady tree must be planted around the shed to provide shadow during summer and to protect chilly wind during winter. Ventilation can also be controlled through fittings of air condition equipments. It is not economic for small scale production but is viable when rabbits are produced commercially in large scale.

Lighting

Light plays important role in rabbit reproduction. Both duration and intensity of light are to be considered in rabbit enclosure. It is well established that buck requires eight hours light out of 24 hours for normal spermatogenesis and sexual activity. On the other hand, 14–16 hours exposure of light in a day favours normal sexual activity and fertilization in doe. The intensity of light throughout the shed should be 30–40 lux. Fluorescent tubes can be used for this purpose.

Noise

Rabbit house must be constructed away from the main road to avoid noise from the running vehicles. Crowded area is also to be avoided. Internal atmosphere of rabbit shed must be calm and quiet particularly for normal copulatory and maternal behaviour.

CONSTRUCTIONAL DETAILS

Floor

Floor should be durable, hard, non-slippery, non-porous, dried, easily cleaned, and free from dampness. Although the floor can be made up of hard soil, moorum/red soil/ kankar, cement concrete, stone slab, and brick-on-edge etc. Out of these materials cement concrete floor is mostly preferred for rabbit shed. Sometimes carbo rundum powder is mixed with the cement mixture to make the floor non-slippery. The surface of cement floor is generally roughened by imprinting the impression of either wire mesh or extended metals. Proper slope towards the gutter is to be given to facilitate drainage of water and urine. If rabbits are reared in cages, pits of 1–2 inches deep is constructed under the cages for the periodical collection of manure/ rabbit wastes. There should be at least three feet distance between two rows of cages and from the wall for the movement of persons working in the shed or inspecting the cages.

Roof

Roof should be strong, durable, fire proof, light weight and should protect from bad climate. It may be made up of reinforced cement concrete (RCC), reinforced brick concrete (RBC), thatch materials, tiles, slates etc. Considering the easy maintenance of internal room temperature of rabbitry either cement concrete or thatch roof is preferred.

Wall

The basic functions of wall of any animal shed are: (*a*) to provide support to the roof, upper structures and fittings of the building, (*b*) to protect the floor from hot, cold including rain, and (*c*) to protect the animals from theft as well as attack from predators. The height of the wall at eaves is generally 8–10 ft. The wall can be made up of mud, brick, stone, sheet, locally available materials etc. The normal width of the wall should be 10–12 cm. Under Indian climate, long axis of a building should be in east to west direction. Generally, coolers or air conditioners are fitted in the south wall and exhausts are in the north for proper air flow within the shed and easy control of internal temperature throughout the year. To facilitate proper ventilation and light within the building either hopper or centre pivoted window with glasses are fixed in the walls. The building must be insect proof.

HOUSING SYSTEMS

Rabbits can be housed both in outdoors and within the building. Outdoor housing is comparatively less expensive and provides better ventilation including light, but, has poor security as well as protection from bad climate than indoor housing system. The housing system mainly depends upon the ambient temperature of the region. The hutch area depends upon breed, size of rabbit and number of animals to be reared or intensity of production. Considering the welfare ample space should be provided to rabbits for their normal activity. The following hutches are generally used to rear rabbit either in domestic purpose or in commercial production.

Wooden/Concrete Cages with Litter

Cage may be made up of either cement concrete or wood.

Soft bedding materials such as wheat straw, paddy straw, wood shavings, rice husk, cutting papers etc. are generally used as litter which should be replaced every week to control parasites and other microorganisms. For one breeding rabbit an ideal dimension of cage is 80–100 cm in length, 60–70 cm in width and 50–60 cm in height. The advantages of this type of housing are economic, readily available and save labour in cleaning with absorbent materials. Maintenance of good hygiene is the major constraint.

Wooden/Concrete Cages Without Litter

Rabbits are reared directly on the cement concrete or wooden plank without any litter. It causes high humidity and bad hygiene within the room leading to poor health of rabbits. As a solution, this housing system is modified to separate rabbits from their excreta as soon as dropped. The rabbits are raised above ground on slatted floor or wire mesh. The wire mesh floor should be constructed in such a way that it will not injure foot pad of rabbits and side by side, droppings will fall through mesh. Normally 1.3 to 1.5 cm space between slats is sufficient enough for droppings to fall through. Various kinds of slats *viz.,* bamboo, wood, plastic and metals are used by the commercial rabbit producers. Out of these, wire mesh is mostly preferred due to its comfort.

Outdoor Colony System

In this system, rabbits are kept outside of a building in a loose colony covered with wire-mesh fencing at about 6–7 ft height. Top of the fence is angled outwards or bent out to avoid entry of predators. The base of fencing wire is fitted deep into the ground to about 2–3 feet to prevent rabbits from burrowing out of fence. Sometimes, a number of discard pipes are put in the fence to fulfil their natural burrowing habit. The required number of nest boxes should be put in the colony. The size of the colony mainly depends upon the number of rabbits to be kept.

Morant Hutch

The hutch is a free-standing unit in which one or more number of rabbits can be kept. Hutch must be comfortable, escape proof, sufficiently large, easy to handle, easy to clean, well constructed of sound materials and so constructed that the rabbits can neither hurt itself nor the hutch. Major G Morant invented a special hutch for rabbit towards the end of nineteenth century which is known as Morant hutch. It has a triangular top with one third space covered by wooden planks and rest by wire mesh. The normal size of this hutch is about 60–80 inches in length, 30–35 inches in width and 30–40 inches in height. Through this housing rabbits can graze directly from the ground. Within the covered area, there is a raised dry platform where they can sleep. Generally, rabbits reared for show purpose or as pet can be kept by this type of light weight hutch. The major advantages are easy handling of cage due to light weight, grazing of rabbit directly from ground and prevent foul smell and automatically fertilize the area. But, large number of rabbits can't be reared efficiently by this system.

Wire Cage

Cage is made up of welded galvanised wire mesh of 16–14 gauge of 1.0 × 2.0 sq. inches size in front, back, top and sides. The bottom should be of 0.5 × 0.5 sq. inches size. It can be made up of either single or in one row or two rows of a particular length so that the cages can be extended in any direction as per requirement. The dimension of a single cage is 0.75 metre in length, 0.75 metre in width and 0.65 metre in height. The cage is generally hanged from the roof of building to avoid attack from rats and other predators.

The height is adjusted in such a way that they can be handled very easily by the workers / stockpersons. The types of cages, used in rabbit housing, are described below.

Flat deck cage

The cages are fixed at one level in a row by hanging them from roof through iron chains or rods. Double row cage as single unit is also available at the market where hay rack is fixed in between the two rows of cages. The cage usually opens from the top. Depending upon the cleaning frequency either shallow or deep pit is prepared just under the cages to collect excreta coming directly from rabbits (Fig. 19). The advantages of this type of cages are: (*a*) easy supervision and handling of rabbits, (*b*) better comfort to animals, (*c*) no requirement of elaborate ventilation, and (*d*) long life of cage. But, the disadvantages are low stocking density per unit area of building and high initial investment cost per rabbit.

Fig. 19: Flat deck cages

Californian cage

In this system, cages are fitted in a row one deck higher than other but not just above. It may be of two or three or multiple tiers according to the requirement (Fig. 20). Lower

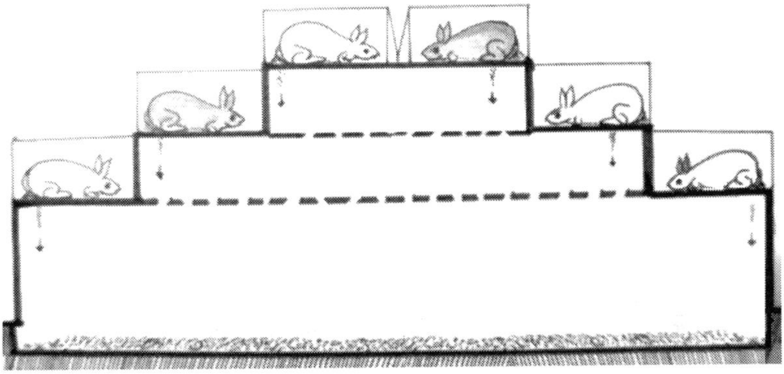

Fig. 20: Californian cage

row of the cages open at the top and higher rows from front. The excreta from each row/tier drops directly to ground just like flat deck system. The major advantages are: (*a*) high stocking density per unit area of housing, (*b*) dropping falls directly to the ground which does not contaminate other rabbits kept in other tiers/steps, (*c*) provides good ventilation, and (*d*) per animal investment is comparatively lower. But, there is a problem of handling and supervision of rabbits kept in upper tiers.

Inclined slope battery cage

The cages are fitted one above another and excretion trays are inclined towards rear side of the cage in such a way that the excreta will come down through tray to the pit on ground from where it can be removed either manually or automatically (Fig. 21). Generally, the cages are opened from the front. The high stocking density per unit area is the main advantage of this type of cage housing. The drawbacks are: (*a*) it is difficult to handle and supervise the rabbit of upper step, (*b*) there is the chance of contamination of lower cages, and (*c*) ventilation and lighting are comparatively poor.

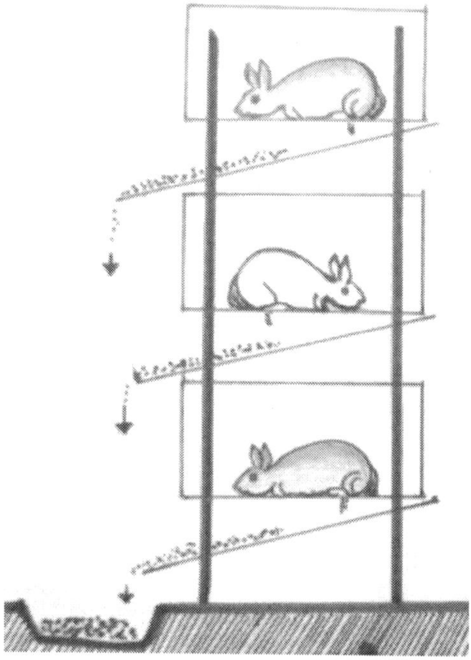

Fig. 21: Inclined slope battery cage

Compact battery cage

This system is almost similar to inclined slope battery system. Here, the waste collection trays are fitted below each cage of each step parallel to the ground, not inclined (Fig. 22).

FEEDING AND WATERING EQUIPMENTS

Each cage must have feeding trough/hopper (Fig. 23), water trough and forage rack. Sizes of troughs and forage rack depend upon the number of rabbits kept in the cage. The troughs and rack must be strong enough to withstand bite of rabbits, easy to fill from outside the cage, easy to clean,

Fig. 22: Compact battery cage

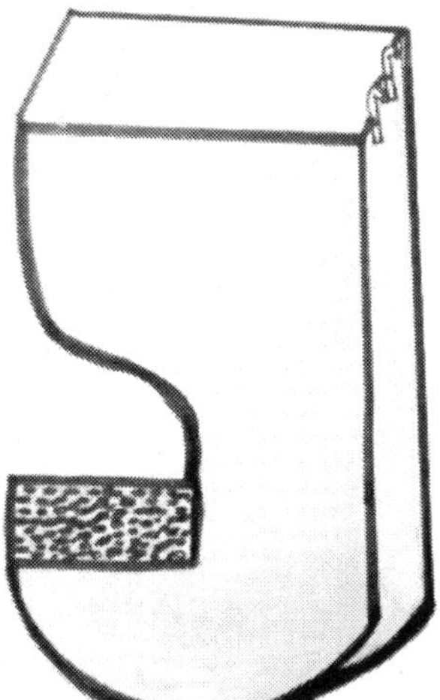

Fig. 23: Feeder for rabbits

disinfect and detachable, and fixed tightly with the cage to protect from wastage and predators. If rabbits are reared in deep litter system, the hopper and rack are to be arranged in such a way that the rabbits cannot overturn them or cannot climb on these.

Rabbits must have constant access of drinking water to maintain the productivity. Water can be supplied by using water bowls, bottle drinkers that hang on the cage, or automatic watering systems. When rabbits are reared under deep litter housing system they have the tendency to soil water. Maintenance of good hygiene is a difficult task then. The design of water dispenser should be such that: (*a*) it will not be easily turned down by the rabbits, (*b*) it should be as permanent as possible, (*c*) wastage is minimum, (*d*) easily cleaned, and (*e*) easily refilled by the workers engaged in rabbit management. If automatic drinkers are used, the problem does not arise. However, various watering systems, used in rabbitry, are as described below.

Inverted Water Bottle Drinker

A long necked bottle is generally inverted over a small trough to avoid contamination (Fig. 24). The size of bottle should be sufficient enough which does not require frequent refilling and worker can check normal water intake daily. It has less investment but more chance of contamination than others.

Automatic Drinker

This system is fitted with every individual cage and connected with the main water storage or supply tank. It saves labour involvement and makes medication easier. On the other hand, initial investment is high and if the storage tank is not maintained properly, there is the high risk of water contamination leads to quick spread of disease.

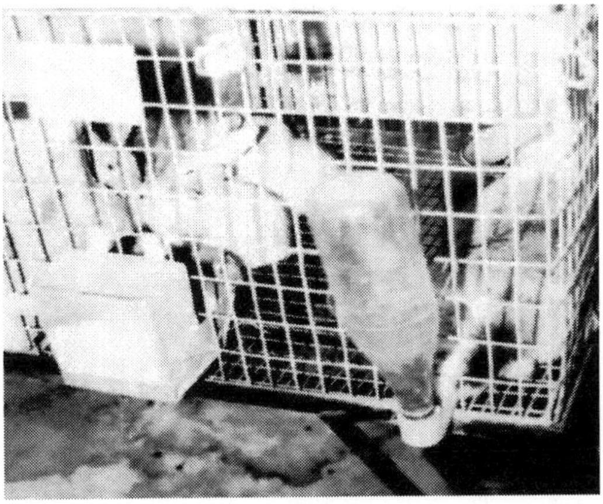

Fig. 24: Cage fitted with feeder and water bottle

Nipple Drinker

Water is supplied in every cage either from centralized storage tank or individual bottle dispenser fitted with nipple (Fig. 25). The advantages are: (*a*) water is more hygienic as there is little or no chance of contamination, (*b*) easy to handle for its smaller size, and easy medication. The problem comes when there is any leakage in the fitted nipple. It wets feeds and bedding materials.

NEST OR KINDLING BOX

Kindling box is one of the most important equipments used in rabbit production. In rabbit, pre-weaning mortality is very high. Utmost care and management is required to save the valuable lives of kits. Kits are reared within this box along with their mother upto weaning. The functions of nest boxes are: (*a*) it protects kits against attack of predators and bad weathers, (*b*) it provides comfortable, healthy and hygienic environment to nurse her kits, (*c*) it maintains constant temperature around 30–35°C in the middle of the nest, (*d*) it favours easy movement of animals, (*e*) it provides damp

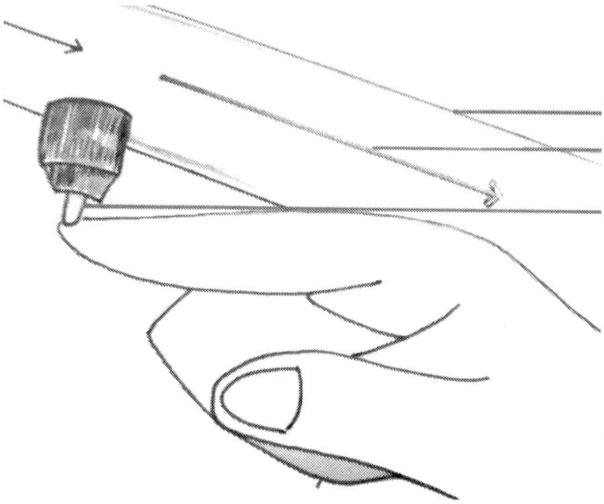

Fig. 25: Nipple drinker for rabbit

free and well ventilated accommodation, and (*f*) it facilitates easy monitoring of young ones by the workers. Half of the nest is made up of wood and half of wires. The dimensions of box for a adult rabbit with her litter should be 1.8 metres in length, 0.9 metre in width and 0.9 metre in height. The wooden part has ventilation aperture with net. Supporting wooden pieces must be fitted outside the wooden cage because, if it remains inside of the box, wood will be chewed by the rabbit due to their gnawing habit. Clean soft straw, dry grasses, coconut fibres, wood scrapings, cutting papers etc. can be used as bedding materials which also maintains the internal temperature (Fig. 26). The depth of bedding materials varies accordingly as per kindling season. There should be one opening in the portion between wire and wooden cage for the movement of animals where there will be the provision of temporary closing of gate. During the day mother can take rest in wire cage and shutter is used to prevent entry her into wooden box. Mothers are allowed to enter into the wooden box once or twice in a day only to nurse her kits. Kits are reared within wooden box until their eyes are opened. After that, kits and mothers can frequently moved to and fro within the whole kindling cage up-to weaning (Fig. 27).

Fig. 26: Kindling box with bedding materials

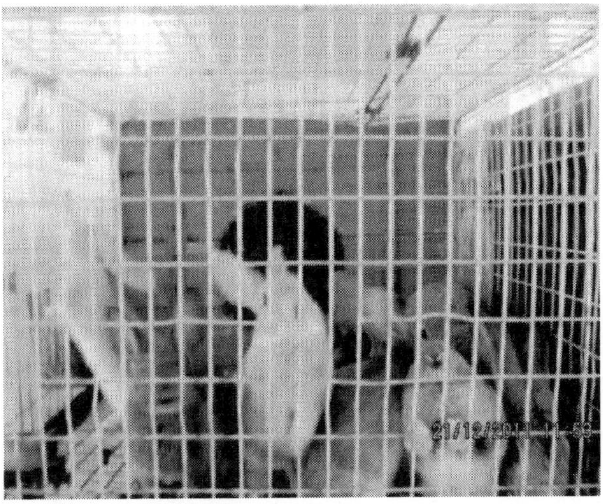

Fig. 27: Kindling box with young rabbits

PASSAGE FOR MOVEMENT

The distance between cages is very important for the movement of persons engaged in the farm. Moreover, the passage is also used for the movement of feed and fodder trolley. The moving passage should be raised than the rest of the floor under the cage to prevent slip arising from the accumulation of urine of rabbit and water from washing of shed. Normally, when rabbits are reared in flat deck cage the cage should be placed at least 3 ft distance from each other and from the wall. The passage must have proper slope to drain washed water from the shed.

FEED STORAGE

There should be the provision for storing concentrate feed/ pellets in rabbit shed. The storage must have the capacity for storing at least 3 months feed in advance. Cement concrete floor is preferred for storing feed materials. The room must be airy, free from damp, rats and other insects. The door must be fitted tightly to keep the vermin out. In some organised rabbitries, even feed is kept in specially constructed tank to avoid any spoilage.

Chapter 6

Health Management

Rabbits are very much sensitive to different kinds of health problems and require an utmost care and management. The incidence of diseases is very much related to the purpose of use, breed, nutritional status, rearing system, hygiene and sanitation, health care etc. Pet rabbits have different needs from wild rabbits. However, it is very essential for the rabbit keepers to know the basic knowledge to differentiate healthy and diseased rabbit. Generally, healthy rabbits have bright colourful and clear eyes, clean nose, smooth and clean ears, shiny and smooth coat, soft and pliable skin, head with normal position and intact claws with healthy foot pads. Rectal temperature, pulse rate and respiration rate vary from 39.0°C–39.5°C, 140–150 beats per minute and 30–65 times per minute, respectively. They are very active, playful and intensely curious in nature. Appetite, thirst and droppings are also normal. Any deviation from these above conditions may be considered as signs of disease or unhealthy rabbit. Then, advice must be taken from any experienced veterinarian for giving proper treatment to rabbit in time. In this chapter, common rabbit diseases with their causative agents, clinical signs and symptoms, diagnosis, treatment, control and preventive measures are briefly discussed.

BACTERIAL DISEASES

Pasteurellosis

The bacterium, *Pasteurella multocida*, is the major infectious agent of rabbits. This is gram negative, non-spore forming short, bipolar and rod shaped bacteria. In rabbit, this disease is characterised by anorexia, weight loss, dehydration, purulent ocular, nasal and vaginal discharge, subcutaneous abscess, pneumonia, incordination, septicaemia and ultimately, death. It is most often transmitted among chronically infected does and their litters or between breeding males and females. Pasteurellosis occurs in a number of clinical forms *viz.*, pneumonia, conjunctivitis, snuffles, abscess, wry neck or otitis media, pyometra, orchitis and septicaemia.

(*a*) ***Pneumonia*:** In domestic rabbit it is very common. Although *Pasteurella multocida* is the predominant bacteria, other bacteria like *Staphylococcus aureus*, *Klebsiella pneumonia*, *Bordetella bronchoseptica* and *Pneumococci* are also involved in this disease. Inadequate ventilation, hygiene of surroundings and other managemental lacunae are the predisposing factors. Rabbit becomes anorectic, listless, dyspneic with sneezing, nasal congestion, and has high temperature (104–105°C). The necropsy finding reveals broncho-pneumonia, pleuritis, pyothorax or pericardial petechie. Chortetracycline, oxytetracycline or slow release penicillin are the drug of choice. The animals should be treated for the longer time. The research finding shows that the organisms present with in pus present in the inaccessible areas where they remain isolated from the therapeutic effect of antibiotics. These microorganisms act as source of re-infection. Due to this reason it is very difficult to complete cure of rabbit.

(*b*) ***Conjunctivitis*:** Both young and adult rabbits are affected. Conjunctivitis can be transmitted either by contact and fomities. It is transmitted when the affected rabbits rub their eyes with the help of front legs and touch cages and bedding materials. The organism may spread from eye to brain as it is directly connected by important nerves. Encephalitis occurs if not treated in time. The affected rabbit shows symptoms like red conjunctiva, swollen eyelids filled with mucopurulent exudates resulting epiphora. Naso-lachrimal duct is frequently clogged with materials and debris resulting spilling over of eyelids by tears and soiling of surrounding hairs. Long term use of topical ophthalmic ointments containing sulphonamide, chloramphenicol or antibiotics with steroid can flush the canal and give satisfactory result. Slow release penicillin is effective against deep seated infections.

(*c*) ***Snuffles*:** This is an acute, sub-acute or chronic inflammation of mucous membrane of air passages and lung which is also known as rhinitis. Sometimes, it is confused with pneumonia and conjunctivitis. From the infected animals along with pasteurella, pseudononous, staphylococcus, streptococcus organisms are also isolated. Low immunity is one of the predisposing factors behind this problem. This disease is characterized by serous, mucous or mucopurulent nasal exudates. Sneezing and coughing are quite common. The animals can be treated with streptopenicillin or sulphaquinoxaline.

(*d*) ***Otitis media*:** This is caused by either *P. multocida* or *Encephalitozoon cuniculi*. The organisms enter through eustachian tube into middle or inner ear, accumulate pus or fluid and cause otitis media or interna. This leads to twisting of its head *i.e.,* 'wry neck' or torticolis. As a result, rabbits are unable to eat and drink. Finally, they die due to dehydration and starvation. Prompt antibiotic therapy may be tried. Prognosis is very poor as drug penetration in the infected area is restricted or difficult leading to mild or no improvement at all.

(*e*) ***Abscess:*** Subcutaneous abscess particularly around eye is very common in rabbits. The organisms responsible are *P. multocida* and *Staphylococcus aureus*. Abscess represents the focal area of infections and may be in single or multiple locations which can grow as large as hen's egg with full of thick toothpaste like puss. The pus can be removed aseptically through surgical interventions. Antibiotic treatment is must to control the spread of infection.

(*f*) ***Genital infection:*** The disease is characterized by serous, mucous and mucopurulent vaginal discharge in doe, enlarge testicles with abscess and orchitis in buck. *Pasteurella multocida* is the main causative organism of this infection in rabbit. The veneral transmission occurs when infected does mate healthy buck or vice versa. Reproductive performance is severely reduced due to this infection. Antibiotic may be tried. But, the prognosis is very poor as diagnosis made ante-mortem is very difficult. Culling and slaughtering of rabbit are the best solution to control the disease.

(*g*) ***Septicaemia:*** This disease is closely associated with and may be considered as sequelae of other form of pasteurellosis. Animals do not show many clinical signs except high body temperature because it occurs in per-acute form. Post-mortem examination reveals the congestion of abdomen, thoracic organ and subcutaneous haemorrhage.

Salmonellosis

Salmonellosis in rabbits is caused mainly by *Salmonella typhimurium* and *S. enteritidis*. *Salmonella pullorum* has also been isolated in some cases. It is a gram negative, aerobic and non-lactose fermenting bacteria.

This disease is characterized by diarrhoea, septicaemia, rapid death and abortion in rabbit. However, the infection is very rare in rabbit. If occurs, young and pregnant are most vulnerable. Generally, organisms are transmitted by ingestion through direct contact with contaminated feeds, water, faeces and litter by rats, mice which carry them. Rabbits show symptoms 4–6 days after picking up the infection. Listlessness, ruffled hair coat, loss of appetite and diarrhoea are the predominant symptoms. In chronic case, animals may not exhibit any symptoms except transient diarrhoea whereas rapid death occurs in acute case. Under necropsy findings enlarged liver, spleen, and haemorrhagic and ulcerative changes in the intestine are common.

Antibiotic like streptomycin may be effective to treat infected rabbits. But, infected animal becomes carrier throughout their lifetime. Therefore, it is advisable to eliminate the colony and restock. Preventive measure can be adopted to keep rabbits away from this disease.

Colibacillosis

Colibacillosis infection in rabbit is caused by bacteria, *Escherichia coli*. It is a gram negative, lactose fermenting, indole positive rod shaped, non-toxin producing and enteropathogenic organism. Weaned rabbits are most susceptible to this disease. Diarrhoea, fever, anorexia and dehydration are the predominant symptoms in rabbits. Edema, pyogranulomatous cellularity of the lamin propria without mucosal ulceration, haemorrhage in the submucosa and thinning of intestinal wall are predominant histopathologic findings. Diagnosis can be made on the basis of isolation of causative organism, symptoms and post mortem findings.

Rabbits should be treated with antibiotics like tetracycline, enrofloxacin, neomycin etc. Fluid therapy as supportive treatment is essential to avoid dehydration. Chlorpromazine injection @1–10 mg/kg body weight intramuscular may help to decrease fluid loss from the secretary diarrhea.

Tyzzer's Disease

Clostridium piliformes (earlier known as *Bacillus pileformes*), a gram negative, pleomorphic, spore forming, filamentous and intracellular bacterium, is responsible for this disease in rabbits. The disease is transmitted through ingestion of spores which remain viable at very low temperature for longer period even more than one year. Weaning rabbits are more susceptible. Acute infection is characterized by profuse watery to mucoid diarrhoea, dehydration and death within 12 to 48 hours after onset of symptoms. The mortality rate is very high. Chronic infection results severe body weight loss, but, rabbit seldom dies.

Lesions in weanling rabbits include edema and haemorrhage of mucosa, submucosa and musculature of intestinal tract. Enlarged liver with multifocal tan to yellow foci of necrosis or haemorrhage of myocardium are common. Extensive mucosal necrosis with a granulomatous cellular mucosal infiltration may occur in the ileum, caecum and proximal colon.

Histopathological examination of liver or caecum stained with silver showing the presence of intracellular rod shaped bacteria is one of the diagnostic tools of this disease. The presence of bacterium can also be found in PCR of faeces, intestinal tissue or liver. To detect antibody in carrier rabbits, ELISA may be helpful. Treatment of infected animals through drug is not economical and very difficult to control. Therefore, it is better to slaughter infected rabbits to eradicate disease from the farm. Strict hygienic measures are the solution to keep rabbits away from this disease. One percent bleach solution can inactivate spore that remains after the fecal material has been washed off soiled cages.

Enterotoxaemia

This disease is induced by the toxins produced by the gram positive bacteria, *Clostridium spiriforme* and *Cl. difficile*. Watery diarrhoea is the predominant symptom of this infection. Sometimes, animal dies suddenly without any previous signs. This disease affects all ages, but, recently weaned rabbits are most susceptible. It causes huge mortality when spreads in the whole rabbitry. Necropsy findings include large, fluid filled edematous caecum with serosal congestion and haemorrhage and watery mucoid faeces in the colon. Histopathologically, there is necrotic erosive or ulcerative typhlitis with swelling and loss of enterocytes and pseudomembrane formation. Heterophils infiltrations are found in mucosa and submucosa. Large bacilli with spores can be observed on the mucosal surface.

Broad spectrum antibiotics may be tried to treat infected animals. Do not use clindamycin and lincomycin in rabbits. Supportive therapy is quite helpful. Control of this disease is very difficult by treating individual animal in the farm. Therefore, adoption of prophylactic measures in rabbitry and elimination of extreme dietary changes, and environmental stressors can check the occurrence.

Tularemia

The disease is caused by *Francisella tularensis*, a gram negative and pleomorphic rod shaped bacterium. It is characterized by depression, anorexia, ataxia and death. Blood sucking arthropods like mosquitoes, squirrel flea, deerfly, woodticks etc. act as vector to spread this disease. Tularemia can be transmitted by direct contact, ingestion, or rarely aerosol. Human beings get infection when they handle infected rabbits. But, the incidence of this infection is very rare in domestic rabbits. Visceral congestion, splenomegaly, consolidated and congested lungs, and multiple pinpoint white foci on the liver and spleen are the typical lesions. No satisfactory treatment is available. Dead animals must be disposed off properly to avoid the spread of disease.

Tryponematosis

This disease is also known as veneral spirichetosis or vent disease or culiculosis or rabbit syphilis. Tryponematosis is caused by *Treponeme paraluis cuniculi,* is a spiral shaped bacterium related to the organism (*Treponema pallidum*) of human syphilis. It is transmitted by direct contact between infected and uninfected rabbits. It is a relatively rare sexually transmitted disease of pet rabbits. Infected rabbits develop raised, crusted and sometimes bleeding ulcers on the external genitalia, around the anus, and on the nose. The popliteal and inguinal lymph node may be enlarged. However, the condition usually disappears after several weeks. Serological test can be done to identify antibody in blood. Infected rabbits is cured by treatment with procaine penicillin @ 40,000 IU I/M b.i.d. for 3–5 days. Breeding of only healthy rabbits and other sanitary measures reduce the incidence of disease.

Pseudo Tuberculosis or Yersiniosis

Pseudo tuberculosis, a zoonotic disease, is caused by the bacterium, *Yersinia Pseudo tuberculosis.* This organism mainly affects rodents *e.g.,* rat, mice, guineapigs etc. Domestic rabbits are usually affected when carrier rodents come in contact with them. Contaminated feed and water are also source of infection. The disease is characterized by anorexia, loss of body weight, depression, diarrhea, emaciation, dyspnoea and death. Lymph node becomes enlarged. There is necrotic foci and lymphoid cell aggregation in the parenchyma of liver, lung and spleen. Infected rabbits may be treated with antibiotics like nitrofurazone or enrofloxacin etc. for longer period. Restriction of rodents to enter rabbitry, use of uninfected feed, water and utensils can check the spread of this disease.

Staphylococcosis

Staphylococcosis is caused by *Staphylococcus aureus*, a gram positive, hemolytic, coagulase-positive coccus. It can be spread through direct contact and infected air. In rabbit, this disease is found in wide range of clinical forms *e.g.*, suppurative infection in any organ or site, subcutaneous abscesses, mastitis with abscess formation, dermatitis, upper respiratory infection with mucopurulent nasal discharge, and septicaemia with depression, anorexia, fever, and death. Out of these, mastitis and abscess formation are commonly reported.

(*a*) ***Mastitis or blue breast*:** Mastitis, a bacterial disease, is not common but is occasionally seen in rabbits. Mastitis is the inflammation of mammary gland which follows injuries of the glands or caked breasts. The affected glands turn bluish in colour due to interruption of blood supply in advance cases. Rabbits show high fever, anorexia and become sensitive to touch the gland. Milk production in lactating doe is severely reduced causing off fed of kits and sometimes leads to death of kits. Early stage treatment is only successful. Penicillin @ 75,000–1,00,000 IU I/M b.i.d. for 3–5 days can be used to treat the infected doe. Reduce the amount of concentrate feed for infected doe, use other milk substitute for kits, disinfect nest box and take hygienic measures to control this disease in the rabbitries. Severely infected animals should be destroyed.

(*b*) ***Abscesses:*** Abscesses may occur subcutaneously or in the viscera. Septicaemic animals show anorexis, hyperthermia and depression. Necropsy finding shows thick walled abscesses filled with purulent exudates. There are diffuse congestion and petechiation of viscera particularly in lung and kidney. Clinicopathological lesions are similar to those of pasteurellosis. Gram staining of exudates can give the differential diagnosis. Culture and antibiotic sensitivities are required for confirmatory diagnosis and treatment. Subcutaneous abscess is found in skin and food pads which should be lanced and flashed with antiseptic solution. Systemic antibiotic is used to prevent secondary infection. Hygienic measures are to be taken to control this infection.

VIRAL DISEASES

Myxomatosis

Myxomatosis is caused by one of the several strains of myxoma virus, a member of the leporipoxvirus group. Myxoma virus infection in cotton tail rabbits (genus-*Sylvilagus*) develops skin tumors whereas in European rabbits (genus-*Oryctolagus)* it causes severe mortality even up to 99%. The acute form of disease is generally characterized by visible lumps and puffiness around the anal, oral, nasal, and genital orifices leading to acute conjunctivitis, skin haemorrhages and convulsions. They show symptoms of listlessness, anorexia, fever, and sometimes there is the problem of secondary bacterial infection which causes pneumonia and purulent inflammation in lungs. Rabbits may survive for 1–2 weeks depending upon the resistance of animals.

In addition to the direct contact with infected materials from infected to health rabbits, mosquitoes, flies, fleas, gnats etc. are the principal vectors to transmit this disease. Contaminated spines of thistles and claws of predatory birds also spread infection. Necropsy findings in *Oryctolagus* rabbit show gross lesions of skin tumors, pronounced cutaneous and subcutaneous oedema on face and body orifices. There may be skin haemorrhages and subserosal petechial and ecchymotic haemorrhages in the stomach and intestines. Depending upon the severity of the infection, degree of visceral congestion varies in different cases. Skin tumors result from initial proliferation of undifferentiated mesenchymal cells which become large stellate cells referred to as 'myxoma' cells. These cells lie in a homogenous matrix of mucinous material. Necrosis may be seen in the centre of this area. Intracytoplsmic inclusions have been observed most prominently in the prickle-cell layer of the epidermis. Diagnosis can be done on the basis of clinical symptoms, pathological lesion, virus isolation, and PCR amplification of viral sequences from tissue specimens. Serological test may also be done.

This infection can be checked by controlling the vectors, adopting quarantine measures during introduction of new animals, isolation of diseased animals and by using other prophylactic measures in the farm. Broad spectrum antibiotics are generally tried to control secondary infection in rabbit. The vaccination can be done at the age of six weeks and repeated yearly, or every six months where the infection is common. The myxomatosis vaccine should not be given within two weeks of a viral haemorrhagic disease (VHD) vaccination.

Fibromatosis

Rabbit fibroma or Shope fibroma virus is a member of leporipoxvirus group and is closely related to myxoma virus, and the squirrel and hare fibroma viruses. Both wild and domestic rabbits are susceptible to this infection.

Although the natural transmission cycle of fibroma virus is not clear, it has been found from the experiment that several species of mosquitoes as well as triatomes, fleas and bedbugs can serve as vectors of this virus. Thus, fibromatosis can be spread through the biting of arthropods. Tumors develop on the legs, feet, muzzle and around the eyes. Size of tumor may be upto 7 cm in diameter and 2 cm in thickness. These are subcutaneous and are not attached to the underlying tissues. The tumors may persist from several months and metastases from the original tumors do not occur. Tumor bearing rabbit does not show other clinical signs. Spontaneous and experimental infections in neonatal domestic rabbits also produce cutaneous and visceral tumors. In cottontail rabbits, initially there is slight thickening of the subcutaneous tissue followed by the development of clearly demarcated soft swelling which may be evident on the sixth day post-inoculation. By the twelfth day tumor reaches its maximum size and persists for months and finally, it regresses. In European rabbit, regression of tumor occurs comparatively at faster rate.

Microscopic lesions show that there is an acute inflammatory reaction followed by localized fibroblastic proliferation accompanied by both mononuclear and polymorphonuclear leucocyte infiltration. Fibroblastic proliferation continues until a distinct tumor is formed consisting of spindle-shaped and polygonal connective tissue cells with abundant cytoplasm. Many cells may have large intracytoplasmic inclusions. Degeneration of the epidermis overlying the tumors may result from pressure ischemia leading to necrosis, sloughing of the epithelium, and the tumor. The diagnosis can be done on the basis of clinical signs and morphological lesions. Vector control is the main way to check the spread of infection.

Papillomatosis

(*a*) ***Rabbit (Shope) papilloma virus infection*:** Shope papillomatosis most frequently occurs as a natural disease of cotton tail rabbits and is caused by the virus of the genus *Papillomavirus* of the *Papillomaviridae* family. The rabbit (Shope) papilloma virus is antigenically distinct from other members of the genus *papillomavirus*.

Mosquito is the main vector of transmission of this disease from feral to domestic rabbits. This disease in cottontail rabbits is characterized by the presence of horny warts usually on the neck, shoulder, or abdomen of the affected rabbit. Some of the infected rabbits show carcinomatous lesions. Papilloma lesions may also be disappeared within six months after infection. In domestic rabbits, papillomata most commonly occur on the eyelids and ears. The growth is well keratinized and the upper surface is irregular and often split. Lower portion is fleshy to touch and pinkish in colour, and lateral surface appears striated.

Tumor has the typical appearance of a papilloma with elongated rete pegs of epithelium surrounding central cores of connective tissue. There is infiltration of inflammatory cell in the dermal layers underlying the tumor. Failure of the lesion to resolve may lead to the development of squamous cell carcinoma. Diagnosis is done on the basis of clinical signs and histopathological examinations. Arthropod vectors control can reduce the spread of infection. Vaccines can be used to control this infection.

(*b*) ***Rabbit oral papilloma virus infection*****:** It is also caused by the papillomavirus of papovavirus group. The domestic rabbits between the ages of 2 and 18 months are the natural host of this disease. Transmission occurs through direct contact of oral secretions containing sloughed epithelial cells from the oral warts. This benign disease is characterized by numerous whitish growths on the underside of the tongue, oral cavity epithelium or gingiva which later become pedunculated and ulcerate. When rabbits become immune the growths regress.

Papilloma shows verrucous epidermal hyperplasia with rete peg formation and hyperkeratosis, and dermal fibroplasias. Intraepithelial viral inclusion bodies are not usually seen on microscopic examination of tumors. It can be diagnosed by gross lesions. Recovered rabbits become resistant to reinfection. Hygienic measures may help to spread the infection.

Pox

Rabbit pox is a very rare disease which is caused by a virus of genus *Orthopoxvirus*, and the family *Poxviridae*. The rabbit pox virus is antigenically very closely related to the vaccinia virus and is distinctly related to myxoma and fibroma viruses. Except few outbreaks since first report in 1930, rabbit pox has not been recognized in wild rabbits. Laboratory rabbits are mainly susceptible. It does not infect human being.

The nasal cavity is the site of primary infection with the virus which appears to multiply in the nasal mucosa and latter in the lymph nodes of respiratory tract and lung, and spleen. After two to three days of getting infection the rabbit becomes febrile with profuse nasal discharge. Popliteal and inguinal lymph nodes become enlarged. After five days, there are the appearance of erythematous and macular rash on skin, and mucous membrane of the oral and nasal cavities which develop into papules. These papules turn into nodules of about one centimeter in diameter which eventually dry up forming superficial crusts. Face and oral cavity become oedematous. Purulent conjunctivitis is frequently occurred. Buck shows the symptoms of orchitis with papule in the sheath and urethra. The doe exhibits similar symptoms in their vulvae. Pregnant infected female may abort. In severe cases, death usually occurs after seven to ten days of infection. A typical nodule consists of a central zone of necrosis surrounded by a zone of mononuclear cell infiltration. There is edema and haemorrhage in the adjacent tissues, perivascular mononuclear and polymorphonuclear cell infiltration and necrosis in lung. Spleen shows congestion and marked distension of the sinuses with mononuclear cells. The cytoplasmic inclusions characteristically associated with pox virus infection are seldom occurred in rabbit pox lesions.

Based on the clinical signs, characteristic gross and microscopic lesions are the bases of diagnosis of this disease. The confirmatory diagnosis may be made by fluorescent antibody test of tissues or by PCR or by other serological tests. Rabbits are treated with antibiotics to check the spread of secondary bacterial infection. Fluid therapy and non-steroidal anti-inflammatory drugs are also found to be beneficial.

Herpesvirus Infection

Herpes virus mainly affects the respiratory tract and genital organs of animals including humans. They can exist in subclinical forms for longer periods, while the carrier animals remain asymptomatic. Normally, stressful conditions activate the dormant virus and cause outbreak in the infected rabbit. Two herpesviruses are observed in domestic rabbits: (*a*) *Herpesvirus sylvilagus*, and (*b*) *Herpesvirus cuniculi*.

(*a*) ***Herpesvirus sylvilagus (Cottontail virus)*:** This virus is isolated from primary kidney cell cultures prepared from the kidneys of apparently normal weaning cottontail rabbits (*Sylvilagus floridanus*). It is transmitted *via* faecal matter, urine, the doe's milk, mating, and even through the air. It is not spread by insects. This virus mainly infects the tonsils. The common clinical signs are restlessness, circling, seizures, coma, and interstitial pneumonia. Infected rabbits show a pronounced leucocytosis, lymphocytosis and spleenomegally. It is very difficult to diagnose this viral infection in live rabbits. However, the virus can be found in mouth, lymphocytes and kidneys of the rabbit.

(*b*) ***Herpesvirus cuniculi (Virus-III)*:** This virus is isolated from domestic rabbits. This virus neither produces a natural disease nor are other species of animals susceptible. Older rabbits act as carrier of this disease. It causes fever and rashes, with the appearance of small blisters on the skin. Histologocally, there is mononuclear infiltration and intranuclear inclusions in the testes, skin, and cornea.

Rabbit Haemorrhagic Disease (RHD) or Viral Haemorrhagic Disease (VHD) or Rabbit Calicivirus Disease (RCD)

Viral haemorrhagic disease (VHD) is a highly contagious disease that affects wild and domestic rabbits of the only species *Oryctolagus cuniculus* and is caused by a Calicivirus, genus *Lagovirus* of the family *Caliciviridae*. The mortlality rate is around 90%. The very young rabbits are resistance to RHD.

The virus can be spread through air, urine, faeces, respiratory secretions, direct contact with an infected animal including fomities *e.g.,* clothing, cages, contaminated food, bedding, feeders, water etc. Flies, fleas and mosquitoes may be the source of spreading infection between rabbits. Depending upon the climatic conditions carriers of the virus may remain infectious from 2–215 days. Most of the outbreaks occur either during winter or spring. Incubation period for RHD virus is between 1–3 days. Majority of the rabbits do not show any external symptoms. Symptomatic rabbit displays the symptoms of fever, squeals and coma leading to death within 12–36 hours. Spasms and bloody discharge from nose may also be seen in some cases. Rabbits show uneasiness, excitement, anorexia, swollen eyelids, paralysis, ocular haemorrhages and paddling in less severe cases. Recovered rabbits from less severe cases usually develop jaundice with weight loss and lethargy followed by diarrhea, constipation and abdominal cramping right before death. Post-mortem lesions include hepatic necrosis and splenomegaly. The liver becomes friable and swollen. The spleen turns into black in colour and is also swollen with round edges. Haemorrhages are seen in many organs and tissues.

This disease may be confused with pasteurellosis, myxomatosis, *E. coli* or *Cl. perfringens* type *E. enterotoxemia* infection, poisoning and heat exhaustion. To confirm this infection in rabbit reverse transcription polymerase chain reaction (RT-PCR), Western blotting, negative-staining immunoelectron microscopy and ELISA test are done in the laboratory from the samples of liver, blood, spleen or other organs of infected animals.

Isolation of infected rabbits, quarantine measures, depopulation, disinfection, surveillance and other prophylactic measures are the main control means to eradicate the disease from rabbitry. Rabbits may be protected using killed virus vaccine with annual booster dose.

FUNGAL DISEASES

Dermatophytosis (Ring Worm Infection or Superficial Mycoses)

Dermatophytosis, more colloquial descriptive term '*ringworm infection*', is mainly caused by fungus *Tricophyton mentagrophytes* in rabbit. Occasionally, *Microsporum canis* is being also isolated from the infected rabbit. Although this infection is not common in domestic rabbit, but, has zoonotic importance. Poor husbandry practices, environmental stressors, poor nutrition, overcrowding, inadequate ventilation etc. are the factors responsible for the occurrence of the disease. The organism can be transferred from the fungi spores from objects or from other animals to rabbits. The infectious fungus can survive for months in the environment and can be transmitted through direct contact and from the air. Infection arises on or about the head, spreads to the legs and feet, and other parts of the body though itching. Younger rabbits are more susceptible than adult.

Ring worm lesions are generally circular and are characterized by an underlying inflammatory reaction. The surface of lesion shows a crusty appearance with patchy alopecia. Histological finding reveals hyperkeratosis, acanthosis, and diffuse infiltration of the dermis with polymorphonuclear leucocytes. Leucocytes and plasma are also seen in the dermis and the hair follicle. There is abscess in the hair follicle with secondary bacterial infection. Diagnosis can be done on the basis of clinical signs, scrapings and identification of spores on hair shaft or mycelia within hair shaft, and culture on Sabaraud or DTM agars. Infected rabbits should be treated topically with fungicidal drugs or griseofulvin (@ 25 mg/kg in aqueous suspension or @ 8.25 gm/kg feed in powder form for two weeks). Isolation of infected rabbit until recovery, disinfection of cages and nest boxes with 10% bleach solutions, proper ventilation in rabbitry and adopting other hygienic measures can keep other rabbits away from this disease.

Aspergillosis (Deep and Systemic Mycoses)

This is a fungal disease and is caused by *Aspergillus flavus* or *A. fumigatus*. It is very rarely occurred in rabbit. Infected animals show respiratory problems accompanied by pulmonary lesions containing hyphae surrounded by eosinophilic 'asteroid bodies'. Histological lesions reveal an extensive interstitial pneumonia with infiltration of mononuclear cells. *Aspergillus fumigatus* can also infect reproductive tract of adult female rabbit leading to abortion in late pregnancy.

PARASITIC DISEASES

Coccidiosis

Coccidiosis, a protozoan disease, is caused by the members of the genus *Eimeria*. Out of nine species of *Eimeria* affecting rabbits, one affects liver, and the other eight affect the intestines. Young rabbits are most prone to disease an older rabbits are resistant.

(*a*) ***Hepatic coccidiosis*:** Hepatic coccidiosis is caused by the protozoa *Eimeria stiedae* which is also known as *Monocystis stiedae, Coccidium oviforme* and *Coccidium cuniculi*. Sporulated oocysts are ingested by the rabbit and sporozoites are liberated by enzymatic reaction. These sporozoites invade the epithelial cells of the bile ducts and begin schizogeny. Large number of merozoites are produced which infect contiguous epithelial cells.

The oocysts develop after gametogony. The unsporulated oocysts pass into the lumens of the bile ducts, and exit the body in the faeces. The sporulation occurs outside the rabbit's body. The disease is transmitted from one to another through ingestion of sporulated oocysts from the environment.

The hepatic coccodiosis has wide range of clinical manifestations. Mild infested rabbit does not show clinical signs. In heavily infested rabbits, there are interruption of normal hepatic function, blockage of the bile ducts, anorexia and debilitation leading to decreased body weight. Diarrhoea can occur at the terminal stages of disease. Liver can be enlarged even upto 20% of the total body weight of rabbits, as compared to about 3.7% in normal body weight. Serum bilirubin is also increased. Necropsy finding shows enlarged liver and yellow exudate containing dilated bile ducts. The gall bladder can also be enlarged and contain exudate. Microscopically, papillomatous hyperplasia of the ducts along with multiple life cycle stages of the organism in the biliary epithelium can be found. Ischemic necrosis of contiguous liver parenchyma is resulted from the compression of adjacent blood vessels by swelling of bile ducts.

Diagnosis can be made by examination of faecal examination, by either concentration or flotation method. Oocysts can also be detected either from the exudate of gall bladder, or microscopic examination of impression smears of the cut surface of the liver.

Coccidiosis is very difficult to eliminate with anticoccidial therapy. Sulfonamides are considered to be the drug of choice. Continuous administration of sulfamerazine sodium @ 0.02% in drinking water can inhibit the development of *E. stiedae*. Sulfaquinoxaline sodium @ 0.05% in the drinking water for four weeks or @ 0.03% in the feed for three weeks is also effective against hepatic coccidiosis. Sulfamethoxine and sulfamezathrene have found to be beneficial in treating this infestation. Control of infestation until development of natural immunity is one strategy to minimize the severity of disease. Side by side, strict hygienic measures must be adopted in the rabbitry.

(*b*) ***Intestinal coccidiosis*:** Intestinal coccidiosis is caused by at least eight species of *Eimeria*. They are *E. intestinalis*, *E. flavescens*, *E. irresidua*, *E. media*, *E. magna*, *E. piriformis*, *E. neoleporis* and *E. perforans*. All of these coccidia are found in different parts of intestine and act as a group rather than individual species. Transmission occurs by ingestion of sporulated oocysts. Then the sporozoites are liberated in the intestinal tract and enter the epithelium of mucosa. The life cycles of *Eimeria* species are similar to those of other coccidia.

Postweanling rabbits are most susceptible to this disease. The infestation may be subclinical and symptoms show wide range of variations from mild to severe depending upon age of animal, degree of infestation, nutritional status, husbandry condition and causative organisms. Severe diarrhoea ranging from mucoid to haemorrhagic, weight loss, anorexia and rough hair coat are the predominant symptoms in young affected rabbits. They show intense thirst and death usually occurs due to dehydration and secondary bacterial infection. Older rabbits may shed coccidial oocysts without any clinical sign. Epithelial cells of small and large intestine are severely damaged resulting in ulceration and a mixed mononuclear and polymorphonuclear exudate. Necrotic areas of the intestinal wall appear as white foci. The location and extent of the lesions depend upon the species of coccidia.

Antemortem diagnosis can be made by examining the faeces by direct smear, flotation or concentration method. Examination of mucosal scrapings and observation of coccidial organisms on histological sections of intestines are the procedures followed for post mortem diagnosis. Recently, polymerase chain reaction (PCR) technology is used as a diagnostic test to detect *Eimeria* spp. in the faeces.

As post-weanling rabbits are mostly affected by intestinal coccidiosis, preventive measures should focus on the preweanling period. An oral vaccine, developed from nonpathogenic strain of *E. magna,* may be tried. Sulfamethoxine @ 75 mg/kg BW in feed for seven days is effective to control this disease. Strict sanitation and elimination of infected animals can also reduce the incidence of infestation.

Encephalitozoonosis

Encephalitozoonosis, a chronic and usually latent disease of rabbit, is caused by the parasite *Encephalitozoon cuniculi.* The disease is characterized by motor paralysis, convulsions, tremor, torticolis, paresis, and coma in rabbits.

Route of transmission is not clear. But, the organisms are found in the urine of affected rabbits. Transmission has occurred by oral administration of urine from infested rabbits. The vertical transmission in the rabbit has also been reported. Lesions in the kidneys of infested animals are common. As a gross lesion in kidney, there are multiple, white pinpoint areas randomly scattered over the surface or small indented grey areas on the cortical surface. Granulomatous nephritis or degrees of interstitial infiltration of lymphocytes and plasma cells with fibrosis and tubular degeneration and dilatation are observed under microscopical findings. Granulomatous encephalitis is the characteristic lesion of this disease in rabbit. The lesion is characterized by a central area of necrosis, surrounded by lymphocytes, plasma cells, microglia, epitheloid cells and sometimes giant cells. Lesions of the spinal cord can also occur. The organisms are often not found in histological sections of the lesions. However, it may be seen floating free in the tubules of the kidney.

As encephalitozoonosis has been considered as a newly recognized disease in immunodeficient human, such persons should take proper medical counsel before handling rabbits. Because, isolates from human beings have been shown to be infectious for rabbits.

Diagnosis is made based on histological examination of tissues and observation of the organism using Giemsa stain or Gram stain or Goodpasture-carbol fuchsin stain. Indirect fluorescence antibody test, complement fixation test, immunoperoxidase test, enzyme immunoassay, and PCR tests can also be done to confirm this disease.

There is no successful treatment of this disease. Control and prevention are also very difficult because antemortem detection of infested animals is nearly impossible. Seropositive rabbits should strictly be culled from the colonies.

Cryptosporidiosis

Cryptosporidiosis in rabbit is caused by the protozoan organism *Cryptosporidium cuniculus*. The organism is found in the intestinal tract of rabbit. Transmission occurs through ingestion of thick-walled sporulated oocysts. Clinical signs are not well described in rabbit. However, dilatation of small intestine is also reported. Histopathological findings include shortened, blunted villi, mild oedema of lamina propria, and dilated lacteals of the ileum.

Helminthiasis

The helminth parasites causing natural infestations are

divided into three major groups *viz.*, nematodes, cestodes, and trematodes.

(*a*) ***Nematodiasis*:** The common nematodes of Trichostrongylidae family which affecting rabbits are *Obeliscoides cuniculi* (rabbit stomach worm), *Graphidium strigosum*, *Nematodirus leporis*, *Longistriata noviberiae, Trichostrongylus calcaratus, T. affinis, T. ransoni*. First two parasites are found in rabbit stomach and remainings are in the rabbit intestine. Nematode under Metastrongylidae family, *Protostongylus boughtoni*, is commonly known as rabbit lung worm. It is found in the bronchi of rabbit. Similarly, rabbit nematodes of Oxyuridae, Trichuridae and Filariidae family are *Passaluris ambiguus* (rabbit pin worm), *Trichuris leporis* (rabbit whip worm) and *Dirofilaria scapiceps*, respectively . *P. ambiguus* and *T. leporis* are found in caecum and large intestine whereas *D. scapiceps* is found under the skin of lumbar region, on the medial side of the tarsus, and in the subcutaneous tissues of the forelegs and hindlegs.

Rabbit shows digestive disturbance due to these infestation. Diagnosis can be made on the basis of recovery of eggs from faeces under the microscope. Broad spectrum nematocidal drugs *e.g.*, albendazole, fenbendazole, mebendazole can be used to control worm burden.

(*b*) ***Cestodiasis*:** Rabbits are definitive and intermediate hosts for a number of tapeworms. The life cycles of these parasites practically precludes accidental infection of domestic rabbits. *Cittotaenia variabilis* and *Raillietina salmoni* are found in small intestine of rabbit (definite host). The Oribatid mites act as intermediate host in former and ant in later.

In case of *Taenia pisiformis* and *T. serialis* infestations, rabbit acts as intermediate host. *T.*

pisiformis infestation is very common in wild rabbits and is found occasionally in domestic rabbits. Most cysticercii are found in rabbit liver in *T. pisiformis* infection. Due to migration of the larvae liver may show white tract. Severely affected rabbit dies due to liver damage. *T. serialis* infestation is occasionally found in wild rabbits but rare in domestic rabbits. However, coeneurus stage of *T. serialis* cestode is developed in the connective tissue of muscle of rabbit. Dog acts as definitive host in both the cases. Diagnosis can be made based on postmortem findings. Cestocidal drugs can be used in affected rabbits. Dogs, ants and mites are to be restricted in the colonies to prevent the spread of disease.

(*c*) ***Trematodiasis*:** The rabbit appears to be natural host for trematode *i.e., Hasstilesia tricolor* infestation. The parasites are found in the anterior part of small intestine in rabbit. They produce eggs that infect a land snail, *Vertigo ventricosa*. Snails liberate cercaria which is thought to encyst on vegetation and becomes metacercaria. Rabbits are infested by ingesting the metacercaria.

Rabbit acts as reservoir host in *Faciola hepatica* or liver fluke infestation. The affected rabbit shows emaciation, unthriftiness, debility, retarded growth and anaemia. Microscopical examination reveals the recovery of egg of liver fluke from the faeces. The infested rabbits are treated with oxyclozanide or rafoxanide drugs. Contaminated grass should not be fed to avoid infestation.

Psoroptic Mange or Ear Mange or Ear Canker or Psoroptic Scabies

This rabbit disease is caused by the nonburrowing obligate mite, *Psoroptes cuniculi*, an arthropod parasite. Lesions are found mainly in the inner surfaces of the external ear. It turns into pruritic and can result in scratching and head shaking. Self mutilation may lead to secondary bacterial infection around head and neck. Egg, larvae, protonymph, and adult stages of mite are present together in the host rabbit. At the initial stage, mites chew the epidermal layer of skin. As the disease progresses the crusty dry bran-like material thickens near one inch diameter which consists of desquamated epithelial cells, inflammatory cells, serum, mites and mite faeces. After removal of exudates the skin becomes red in colour and moist. The ear becomes malodorous.

The histological appearance of otoacariasis consists of a chronic inflammatory response with hypertrophy of the Malpighian layer, parakeratosis of horney layer, and epithelial sloughing. Diagnosis is made by identifying the mites from the crusty exudates of affected ear with the unaided eye or with an otoscope.

Before treatment, affected ears are cleaned to remove dry accumulated crusts. Mineral oil with or without acaricide in the ear canal can kill the mites. Subcutaneous or intramuscular injection of Ivermectin @ 400–440 µg/kg body weight is very effective to treat ear mange. Infested rabbit should be isolated to prevent the spread this disease.

Fur Mite Infestation

(*a*) ***Cheyletiella* spp.**: *Cheyletiella parasitovorax*, *C. takahasii*, *C. ochotonae* and *C. johnsoni* are known as fur mites of rabbit. These are non-burrowing mites live in keratin layer of the epidermis. Mostly they are found in the area over scapulae in rabbits. The rabbit may be partially alopecic and skin may have a grey-white scale. Due to the nonburrowing nature of mites, affected rabbits do not scratch and there is no evidence of pruritis. Lesions are characterized by mild dermatitis, hyperkeratitis, and an inflammatory cell infiltrate. Mite can be identified under dissecting microscope. They can be isolated from skin scraping or brushing fur in the affecting area. Topical acaricides or ivermectin can be used to treat the infested rabbit.

(*b*) ***Listrophorus gibbus***: *Listrophorus gibbus* is also commonly known as fur mite of rabbit. *L. gibbus* is a small and noninvasive obligate mite present at low to moderate incidence in domestic rabbits. They are primarily found on the back and abdomen of rabbits and are nonpathgenic. Diagnosis, treatment and control are similar to *Cheyletiella* mites.

SARCOPTIC MANGE

Sarcoptic mange in rabbit is caused by a burrowing mite, *Sarcoptes scabiei* or *S. scabiei* var. *cuniculi*. This mite is commonly known as itch or scab mite of rabbit. *Notoedres cati* or *N. minor cuniculi* is similar to sarcoptic mite in terms of morphology, life cycle, and public health importance.

These mites burrow the skin to ingest epithelial cells and suck lymph. It causes intense pruritic dermatitis, hair loss and abrasions. Serous encrustations on the affected area of skin and secondary bacterial infection are commonly observed. Lesions are found on the face, neck and ear of rabbits. Severely affected rabbits show general debility, emaciation and death occurs within weeks. Diagnosis can be made by identifying mites from the affected area. Subcutaneous ivermectin injection @ 100 µg/kg body weight is effective to eliminate infestation from rabbits.

DISEASE CAUSED BY LICE

Haemodipsus ventricosus is a most common blood sucking louse in rabbit. *H. setoni* has been reported only from wild rabbits. In severe infestations, rabbit becomes week, emaciated and anaemic. There are thinning and ruffled appearance to the fur with skin inflammation. Pruritic condition may develop and secondary bacterial infection may occur. When the pelt is examined with a dissecting microscope or a hand lens, nits and adult anopleurid louse may be observed on the hair. Infected rabbit can be treated with pyrethrin products, silica gel acaricides or ivermectin (0.2–0.4 mg/kg subcutaneous) at 10 day intervals or 2 successive treatments. Louse spends its entire life cycle on the rabbit with little horizontal transmission. Therefore, isolation can be an effective means of control while treating the infested rabbit.

DISEASE CAUSED BY FLEAS

The common name of Cuterebrid flies is rodent warble flies. Cuterebriasis occurs frequently in wild rabbits, but, in domestic rabbits are infected when reared outdoors. Peak period of infestation occurs during the summer and early fall after which the larvae drop to the ground and pupate. The characteristic lesions include single or multiple large subcutaneous swelling containing encysted larvae with a fistula in the centre. Secondary bacterial infection may complicate the disease. The lesions are treated after removing the larva and flushing the wound, or by surgical resection of the wound. Rabbits are to be shifted from outdoor to indoor to prevent this disease.

MISCELLANEOUS DISEASES

Traumatic Vertebral Fracture or Traumatic Vertebral Dislocation or Hind Quarter Paralysis or Broken Back

This disease is common in domestic rabbit. It arises as a consequence of mechanical damage to the spinal cord. The problem occurs suddenly due to the struggling or inadequate support of the hindquarters during handling of animals. The animal is found paralyzed in the cage. The common site of fracture is the L_7 lumbar vertebral body and its caudal articular process. Fracture is more common than dislocation in this area. The animal shows the symptoms of posterior paralysis or peresis, loss of skin sensation, and loss of motor control of anal sphincter and urinary bladder. In advanced cases, there is overfilling of the bladder, elevated blood urea nitrogen, uremia, decubitus ulcers and staining of the peritoneum with liquid faeces.

Diagnosis is made on the basis of clinical signs, neurological examination and, or radiography. Supportive therapy includes regular expression of the urinary bladder and prevention and treatment of decubital ulcers. Corticosteroid and diuretic therapy may be effective for cases of subluxation with spinal oedema. Surgical correction and fixation are not recommended because of the fragility of the bones. Considering the severity of infection, euthenesia can be recommended.

Trochobezoars or Hair Balls

Rabbits ingest hair by licking or pulling their own fur or cage mates resulting accumulation of hair in the stomach in the form of oval masses, called trichobezoars or hair balls. Earlier this condition was thought to be due to excessive ingestion of hair, but now is considered as a normal finding in rabbit. Practically, reduced gut motility causes fluid loss from stomach and dehydration of stomach content leading to the stagnation of the mass. The reason for the ingestion of the hair is not clearly known. However, insufficient fibre content in the diet plays important role and supply of good quality hay/fibrous green fodder can significantly reduce the incidence. Rabbit stops eating and drinking due to partial or complete blockage of stomach and gastrointestinal tract motility slows down. Further, if the rabbit is fed high protein or carbohydrate diet, the situation becomes problematic because it changes caecal pH and growth of bacterial flora. Ultimately, *Clostridium spiriformes* grows in the tract which can result in death due to the production of iota toxins.

Common clinical findings in rabbits are anorexia, dehydration, depression, weight loss, abdominal pain, lack of excretion of faeces, and sudden death due to gastric or intestinal perforation. To diagnose the problem, hairball can be palpated in live animals but contrast radiography may also be used. Treatment of trichobezoar is often unsuccessful. It should be centered on the correction of dehydration and re-establishment of normal gut function. Mineral oil @ 10 ml per day orally can help the passing of ingesta through the gut. Fresh pineapple juice @ 10–15 ml orally once or twice daily for at least five days may provide sufficient energy source to the animal. Force feeding of pulverized food in water or yogurt also can help to improve the appetite. Analgesia is given to control abdominal pain. Early surgical intervention is important in such cases, as other, subsequent metabolic abnormalities may quickly increase the surgical risk to the animal.

Mucoid Enteropathy

Mucoid enteropathy, a sub-acute enteric disease with high mortality, is also known as mucoid enteritis, mucoid diarrhea, enteritis, enteritis complex, bloat, caecal dysbiosis, scours, and hypoamylasemia.

Diet, intestinal flora and the shift from neonatal to adolescent digestive physiology are the predisposing factors of this disease. Low fibre (<10%) in the diet, bacteria *viz., Escherichia coli, Clostridia welchii type A, C. perfringens,* and Coccidiosis are also associated with the higher incidence of mucoid enteropathy in rabbit.

Clinical signs include anorexia, polydipsia, a subnormal body temperature (99–102°C), rough hair coat, mucoid to liquid, tan diarrhea with perineal staining, and abdominal distension with gas and fluid-filled intestines. Grinding of teeth and squinting of eyes are also observed in rabbits. There is increase in packed cell volume, erythrocyte count and haemoglobin concentration in the blood of affected rabbit. Death occurs in 2–4 days. Common pathological findings are distended fluid and gas filled stomach, watery duodenal and jejunal contents, pasty contents in ileum, impacted caecum and gelatinous mucous in colon. Goblet cell hyperplasia without any inflammatory change in duodenum, jejunum, and ileum is a characteristic sign of mucoid enteropathy although ileum contains moderate number of goblet cells. Goblet cell hyperplasia is also occasionally observed in gall bladder, bile ducts, pancreatic ducts, and in the epithelium of trachea.

Diagnosis is done on the basis of clinical signs, gross lesions and histopathology. Feeding of good quality hay to improve crude fibre in the diet, fluid therapy to control dehydration, and cholestyramine to absorb toxins are recommended for the treatment of affected rabbit.

Shock Disease

Shock disease syndrome in rabbit is characterized by severe hypoglycemia and sudden death. This syndrome was first observed in showshoe hares. Many of the trapped rabbits, particularly those were trapped for several times, were found dead in the trap with no evidence of infectious diseases or any kind of trauma. Similar deaths were observed in showshoe hares taken to the laboratory for the research purpose. The syndrome is also found in fatty rabbits or rabbits that have been reared on a rich pellet diet and have undergone a stressful event. Sudden stressful and frightening situation in rabbit release some chemicals (neurotransmitters) from the pituitary gland which release epinephrine or adrenaline. This adrenaline stimulation quickly exhausts the meager glycogen reserve with hypoglycemia and death occurs within a short period. In rabbit, the important clinical signs are convulsions, opisthotonos, dyspnoea, teeth grinding, fixation of eyes and death. Blood sugar level of affected rabbit comes down to 15 to 17 mg% compared to 115 mg% in clinically normal rabbits. Liver glycogen level is depleted to 0.02-0.15% from 5.5% of normal rabbit. The animals are suffered from leucopenia with an absolute lymphopenia. Post mortem finding does not show specific changes except the liver of affected rabbit becomes friable and undergoes fatty changes. Diagnosis is based on clinical signs, low blood sugar levels and typical changes of liver. This syndrome can be prevented to avoid stressful and frightening situation, and obese condition of rabbit.

Sore Hock or Ulcerative Pododermatitis

This is a chronic ulcerative granulomatous dermatitis of the planter surface of the metatarsal region. Although the disease is named as 'sore hock', the volar surface of the metacarpal-phalangeal region is least affected. The disease is caused by the pressure necrosis of the skin resulting from the bearing of heavy body weight on wire-floor housing, foot stomping, or having thin planter fur pads. Lesions include circumscribed ulcerated area in the skin covered by a dry crusty scab varying in different sizes. Secondary bacterial infection with *Staphylococcus aureus* of the dermis adjacent to the ulcerated epithelium may occur and abscesses form under the crusty debris. Anorexia, weight loss, and stilted movement are the predominant clinical signs. Osteomyelitis and sometimes death may occur in neglected rabbits. Rabbits are treated with topical antiseptic solutions after cleaning the debris of wound. Systemic antibiotic may be helpful to control secondary bacterial infection. Side by side, solid bottom cage with dry and soft bedding should be used to minimize trauma.

Heat Prostration or Heat Stroke

Rabbits are very sensitive to heat stress when ambient temperature reaches above 95° F during summer. It may be due to the large ratio of surface area to body weight and insulation resulting from fur. The clinical signs in rabbit are tachypnea, cyanosis, prostration, and blood-tinged fluid on nose and mouth. The affected rabbit should immediately be immersed in cool water or covered with water soaked cloth. The situation can be prevented by maintain constant temperature and ventilation within the shed through using fans, foggers, water sprays and providing air conditioners. Long-haired and wooly rabbits should be clipped at the starting of the summer season. Feeding is also to be regulated as per environmental temperature.

Moist Dermatitis or Slobbers or Wet Dewlap or Ptyalism

This disease is resulting from the chronic wetness of the skin of the chin, mandibular space, and cervical ventrum predisposes to secondary bacterial invasion of these tissues. The chronic wetting of the skin is resulted from the drooling from chronic dental diseases like malocclusion, drinking from crocks of water container, and cold and damp bedding. Necrobacillosis is most commonly seen in moist dermatitis infection. Gross pathological lesions include inflammation, alopecia, ulceration and necrosis of the affected skin. Irregular patches of ulceration, coagulative necrosis, abscessation and colonization of the dermis layers with bacteria are seen under microscope. The incidence can be decreased to rectify the cause of chronic wetting. Rabbits may be treated with broad spectrum antibiotics.

Malocclusion or Mandibular Prognathism or Buck Teeth or Walrus Teeth

This is the most common inherited disease of domestic rabbits. Hereditary malocclusion is apparent by 3 weeks of life and is due to autosomal recessive trait with incomplete penetrance. As rabbit incisor teeth grow 4 inches per year, it is very necessary to clip them every 6 weeks. Here, rabbit's teeth do not align and continue growing leading to difficulty to shut mouth. When the teeth grow long enough, the rabbits are unable to open its jaws or mouth wide enough to eat. Sometimes, this condition causes infection and sores. Malocclusion mainly occurs in front teeth, but, in some cases, the back, or cheek teeth have also been affected. Clinical signs include anorexia and weight loss. Overgrown teeth should be trimmed every month. As the problem is hereditary, affected animals should not be bred for further generation.

Splay Leg

A number of disorders, characterized by complete abduction of one or more legs and the inability to assume a normal standing position, are generally described by the term 'splay leg'. It is a familial or polygenic condition with varying clinical manifestations. Young rabbits of 3–4 weeks of age are mainly affected. Affected rabbits face difficulty in accessing food and water due to inability of leg movements. Typical clinical signs are secondary to femoral endotorsion, with a shallow acetabulum but without luxation of the femur at the hip. The semitendinosus muscle of affected rabbits is abnormal, with smaller fibres and abnormal mitochondria.

PREVENTION AND CONTROL OF RABBIT DISEASES

The proverb '*One ounce of prevention is worth a pound of cure*' should be the main objective of herd health programme. Moreover, rabbits are most sensitive to any kind of diseases. Once rabbits are infected, show very little power of recovery through treatment. If the treatment goes for the longer time, there is more chance to spread the disease to other stocks. Early separation of disease animals from healthy stocks and prompt treatment may save the valuable life of rabbits. Sometimes, humanae slaughtering may also be required to control the disease outbreak. Therefore, a holistic approach on prophylactic measures must be adopted to minimize the incidence of disease which is pointed below.

- Daily inspection of rabbits by trained/experienced person is very important. Disease animals should be immediately isolated and treated. They should only be mixed with healthy stock after complete cure.
- Cages, carrying boxes, crates, utensils etc. should be sterilized before use. House must also be cleaned and disinfected.
- Provide comfortable environment *i.e.*, 15–20°C temperature, 70% relative humidity and adequate ventilation within the shed. House should be free from dust and noise.
- Use foot bath or foot pad with disinfectant before entering in the shed. Maintain strict hygiene within and surrounding the shed.
- Sterilized feed should be used for rabbits. Avoid infected feeds and fodders for feeding of animals.
- Water must be fresh, clean and free from microorganisms available for all the times.
- Do not use infected bucks and does for breeding of rabbits. Check that there is no relation among breeding males and females at least for three generations to avoid inbreeding.
- Outsiders should never be allowed to touch rabbits and cages directly as man is the most dangerous permanent vector of many diseases. Control dogs, vermins, rats and prevent other animals from coming into contact with rabbits, cages, feedstuffs, utensils etc.
- Use quarantine shed for the newly purchased rabbits.

Chapter 7

Economics of Rabbit Farming

In the recent years, commercial rabbit farming has become a lucrative livestock enterprise due to its low initial investment, easy handling, low space requirement compare to other domesticated livestock species and high cash return within the short period of time. Moreover, rabbit meat is lean, white with high protein, low fat as well as cholesterol content and wool quality is quite better than most sheep breeds. Rabbits can provide a valuable additional source of income under backyard farming particularly in the temperate climate of hilly region where employment opportunities are limited. Under small scale production, it does not require expensive housing input and other investments. In case of large scale commercial rabbit production, there are requirements of land development, construction of sheds, equipments, purchasing of breeding parent stock, purchasing of feed items etc. which involves high capital investment. Under this situation, the borrowers can submit a comprehensive techno-economic feasibility report of farming for getting financial assistance from National Bank of Agriculture and Rural Development (NABARD) or nearest any other banks. The report normally should includes information on area, land, availability of breeding stock, availability of feed and water, veterinary aids, availability and cost of daily labourers, experience of farmer, marketing facilities, input costs, output costs, annual expenditure, profit and loss statement, repayment period, and type of assistance available from State or Central Government Institutions/Agencies or any other Non-Government Organizations.

Before preparing any techno-economic feasibility report, it is very essential to know the following basic technical information/norms of meat and wool producing rabbits.

- Breed available for meat - Soviet Chinchilla, New Zealand White, Grey Giant, White Giant etc.
- Breed available for wool - Angora
- Breeding age of animals - 5 to 6 months
- Source of animal- Disease free and unrelated breeding/ parent stock at least for three generations
- Ratio of male to female - 1:5
- Ratio of male and female kit to be born - 50:50
- Gestation period - About 30 days
- Kindling percent - 80 to 90%
- Number of kindling per year per doe - 4 to 6 depending upon facilities available
- Average number of kits born per litter - 4 to 7 depending upon breed available
- Weaning of kits - 42 days
- Rebreeding of doe- Immediate after weaning
- Cage size (adult) - 360 sq. inches
- Number of young (meat rabbits after weaning) to be kept per cage - 4
- Number of adult meat/wool rabbit to be kept per cage - 1
- Kit mortality - About 10%
- Young/adult mortality - About 5%
- Concentrate (pellet) requirement for adult - 120 gm/ day and for young 50 to 100 gm/day

- Hay requirement for adult - 80 gm/day and for young 40 to 60 gm/day
- If hay is not available, leguminous green fodder requirement for adult - 250 to 300 g/day and for young 150 to 200 gm/day
- Cost of breeding stock - Rs. 500/- per animal
- Cost of concentrate feed (pellet) - Rs. 1200/- per quintal
- Cost of hay - Rs. 120/- per quintal
- Cost of green fodder - Rs. 100/- per quintal
- Cost of daily use equipments - Rs. 5000 to 10000/- per year depending upon the number of rabbits.
- Charges of water/electricity/medicine etc. - Rs. 10 per rabbit
- Cost of construction of shed - Rs. 400/- per sq. m.
- Cost of wire cage - Rs. 100/- per cage
- Cost of kindling cage - Rs. 200/- per cage
- Depreciation - 10%
- Market age - 12 weeks
- Market weight - 2.25 to 2.50 kg
- Dressing percentage - 56%
- Wool production from adult - 700 gm per year and young - 250 gm per year
- Value of meat - Rs. 150/- per kg
- Value of wool - Rs. 1000/- per kg
- Value of skin - Rs. 20/- per kg
- Value of manure - Rs. 40/- per adult meat/wool rabbit and Rs. 10/- per young meat/wool rabbit
- Value of empty feed gunny/plastic bags - Rs 10/- per bag
- Bank loan interest - 12.5%

- Margin money - 10%
- Women and other unemployed persons of a family are to be engaged in the project

Project 1: Prepare a techno-economic feasibility report of a broiler rabbit farm starts with 72 Soviet Chinchilla breed considering the above technical norms. Maintain 96 parent stocks with proper male female ratio at the end of year for production in the next year.

Calculation:

Number of females parent stock to be purchased @1:5 ratio
= 60

Number of males parent stock to be purchased @1:5 ratio
= 12

Number of does produce kits = 90 (kindling %) x 60
= 54

Number of kits born after kindling of 1st batch = 54 x 6
= 324

Number of dead kits after kindling of 1st batch = 324 x 0.10
= 33

Number of survived kits after kindling of 1st batch
= 324 – 33 = 291

Cage requirement for young kits after weaning of 1st batch
= 291/4 (4 kits /cage) = 73

Number of survived young animals from 1st batch
= [291 – (291 × 0.05)]
= 291 – 15 = 276

Total number of survived young from 4 batches of kindling
= 276 × 4 = 1104

As four kindlings will be taken from each doe in one year and young will be marketed for 84 days, one set (73 in nos.) cage will be sufficient enough for keeping of young kits upto market *i.e.,* 84 days of age in addition to 72 cages for parents, one set (54 in nos.) nest/kindling boxes, 15 surplus cages for young and 6 surplus kindling boxes. The production plan of meat rabbits is shown in Fig. 28.

I. ***Capital expenditure*:**

A. **Shed and cages:**

1. Cost of construction of rabbit shed (Kachha stone masonary and thatch roof) of size 6 × 12 m × 5 m = 360 sq.m. @ Rs. 400 per sq. m. = Rs. 1,44,000/-

2. *Cost of cage:*

   a. All-wire cage (0.5 m × 0.5 m) = 160 cages @ Rs. 100/- per cage = Rs. 16,000/-

   b. Kindling/nest box = 60 @ Rs. 200/- per cage = Rs. 12,000/-

Subtotal (1 + 2) = Rs. 1,72,000/-

B. **Cost of daily use article** (Bucket, scissors, wire brushes, feeders, waterers etc.) = Rs. 10,000/-

Subtotal (A + B) = Rs. 1,82,000/-

C. **Cost of rabbits** 72 rabbits @ Rs. 500/- per rabbit = Rs. 36,000/-

Total (A + B + C) = Rs. 2,18,000/-

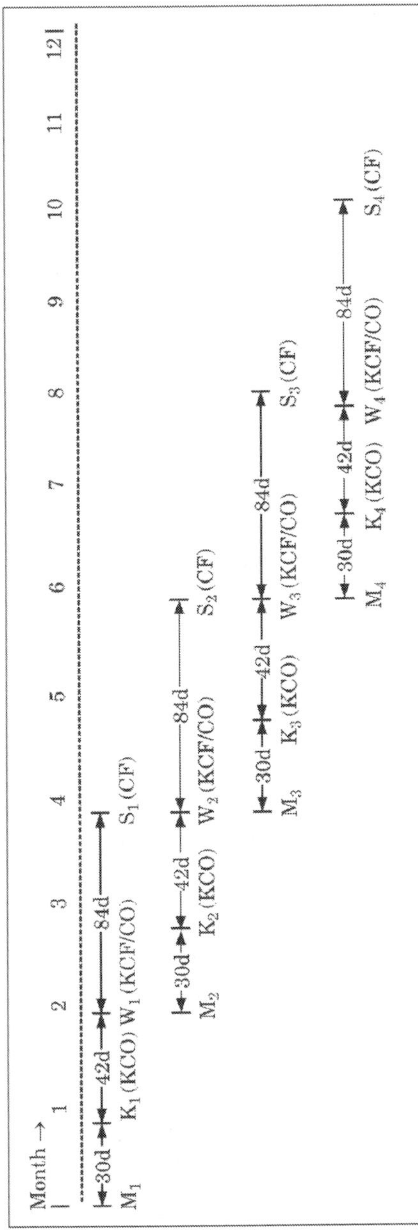

M = Mating; K = Kindling; W = Weaning; S = Sell; 1, 2, 3 & 4 = Sets of production; KCO = Kindling cage occupied; KCF = Kindling cage free; CO = Cage occupied by young; CF = Cage free and d = Days

Fig. 28: Production plan of meat (Soviet Chinchilla) rabbits (4 kindlings per year)

II. ***Recurring expenditure (for 12 months):***

A. **Cost of concentrate:**

1. *Adult:* 72 × 120 × 365 gms
   = 31.536 qtls. @ Rs. 1200/-
   = Rs. 37,843/-
2. *Young:* 1104 × 50 × (84 – 30*) gms
   = 29.808 qtls. @ Rs. 1200/-
   = Rs. 35,770/-

B. **Cost of hay**

1. *Adult:* 72 × 80 × 365 gms
   = 21.024 qtls. @ Rs. 120/-
   = Rs. 2,523/-
2. *Young:* 1104 × 40 × (84 – 30*) gms
   = 23.846 qtls. @ Rs. 120/-
   = Rs. 2,862/-

C. **Miscellaneous cost**

Water/electricity/medicine etc.

= 1176 × Rs. 10/- = Rs. 11,760/-

Total (A + B + C) = Rs. 90,758/-

Total expenditure (capital and recurring cost)

= Rs. (2,18,000 + 90,758)/- = Rs. 3,08,758/-

III. ***Income from first year:***

A. **Sale value of meat:**

1080 (1176 – 96) × 1.3 kg × Rs. 150/- = Rs. 2,10,600/-

* Kits are taken their mother's milk for 30 days after birth

B. **Sale value of skin of rabbits:**

1080 × Rs. 30/- = Rs. 32,400/-

C. **Sale value of manure:**

1080 × Rs. 10/- = Rs. 10,800/-

D. **Sale of empty gunny bags:**

123 bags × Rs. 10/- = Rs. 1,230/-

Total (A + B + C + D) = Rs. 2,55,030/-

IV. ***Cost of production:***

A. **Recurring cost:** = Rs. 90,758/-

B. **Depreciation cost:**

Rs. 1,82,000/- × 10% = Rs. 18,200/-

C. **Margin money:**

Rs. 3,08,758/- × 10% = Rs. 30,876/-

D. **Interest of bank loan:**

Rs. (3,08,758 – 30,876)/-

= Rs. 2,77,882/- × 12.5% = Rs. 34,735/-

Total (A + B + C + D) = Rs. 1,74,569/-

V. ***Net profit from first year*:**

A. **Total income:** = Rs. 2,55,030/-

B. **Total cost of production:** = Rs. 1,74,569/-

Net profit (A – B) = Rs. 80,461/-

Project 2: Prepare a techno-economic feasibility report of a broiler rabbit farm starts with 12 German Angora breed considering the above technical norms. Maintain 36 parent stocks with proper male female ratio at the end of year for production in the next year.

Calculation:

Number of females parent stock to be purchased @ 1:5 ratio
= 10

Number of males parent stock to be purchased @ 1:5 ratio
= 2

Number of does produce kits = 90 (kindling %) × 10
= 9

Number of kits born after kindling of 1st batch

= 9 × 4 = 36

Number of dead kits after kindling of 1st batch

= 36 × 0.10 = 4

Number of survived kits after kindling of 1st batch

= 36 – 4 = 32

Cage requirement for young kits after weaning of 1st batch
= 32

Number of survived young animals from 1st batch

= [32 × (32 × 0.05)]

= 32 × 2 = 30

Total number of survived young from kindling of 4 batches

= 30 × 4 = 120

Number of cages to be required: 120 (young) + 12 (adult) + 10 (surplus)
= 142

Number of kindling boxes to be required: 9 +1 (surplus)
= 10

The production plan of wool rabbits is shown in Fig. 29.

I. ***Capital expenditure:***

A. **Shed and cages:**

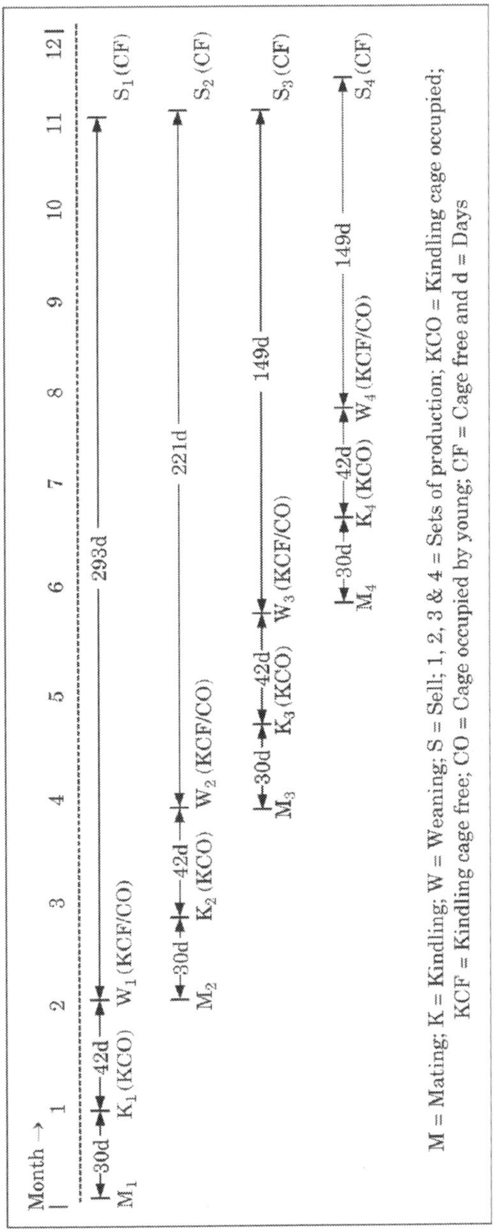

Fig. 29: Production plan of wool (Angora) rabbits (4 kindlings per year)

1. Cost of construction of rabbit shed (Kachha stone masonary and thatch roof) of size 12 m × 5 m = 60 sq.m. @ Rs. 400 per sq. m. = Rs. 24,000/-
2. Cost of cage:
   a. All-wire cage (0.5 m × 0.5 m): 142 cages @ Rs. 100/- per cage = Rs. 14,200/-
   b. Kindling/nest box: 10 @ Rs. 200/- per cage = Rs. 2,000/-

Subtotal (1 + 2) = Rs. 40,200/-

B. **Cost of daily use article** (Bucket, scissors, wire brushes, feeders, waterers etc.) = Rs. 5,000/-

Subtotal (A + B) = Rs. 45,200/-

C. **Cost of rabbits** -12 rabbits @ Rs. 500/- per rabbit = Rs. 6,000/-

Total (A + B + C) = Rs. 51,200/-

II. ***Recurring expenditure (for 12 months)*:**

A. **Cost of concentrate**

1. ***Adult:*** 12 × 120 × 365 gms = 5.256 qtls. @ Rs. 1200/- = Rs. 6,307/-
2. ***Young:*** *First batch:*

30 × 305* days × 120 gm
= 10.98 qtls. @ Rs 1200/- = Rs. 13,176/-

Second batch:
30 × 235* × 120 gm
= 8.46 qtls. @ Rs 1200/- = Rs. 10,152/-

Third batch:
30 × 160* × 100 gm
= 4.80 qtls. @ Rs 1200/- = Rs. 5,760/-

Fourth batch:
30 × 120* × 50 gm
= 1.80 qtls. @ Rs 1200/- = Rs. 2,160/-

B. **Cost of hay**

1. ***Adult:*** 12 × 80 × 365 gms
= 3.504 qtls. @ Rs. 120/- = Rs. 421/-

2. ***Young:*** *First batch*
30 × 305* days × 80 gm
= 7.32 qtls. @ Rs 120/- = Rs. 878/-

Second batch
30 × 235* × 80 gm
= 5.64 qtls. @ Rs 120/- = Rs. 677/-

Third batch
30 × 160* × 60 gm
= 2.88 qtls. @ Rs 120/- = Rs. 346/-

Fourth batch
30 × 120* × 40 gm
= 1.44 qtls. @ Rs 120/- = Rs. 173/-

C. **Miscellaneous cost**

Water/electricity/medicine etc.

---

* Kits are taken their mother's milk for 30 days after birth

= 132 × Rs. 10/- = Rs. 1,320/-

Total (A + B + C) = Rs. 41,370/-

Total expenditure (capital and recurring cost)
= Rs. (51,200 + 41,370)/- = Rs. 92,570/-

III. ***Income from first year*:**

A. **Sale value of wool:** (12 × 700 + 120 × 250) gm
= 38.4 kg @ Rs. 1000/- per kg = Rs. 38,400/-

B. **Sale value of manure:** (12 × 40 + 120 × 10)
= Rs. 1,680/-

C. **Sale of empty gunny bags:** (63 bags × Rs. 10/-)
= Rs. 630/-

D. **Sale of rabbits:** (96 × Rs. 500/-) = Rs. 48,000/-

Total (A + B + C + D) = Rs. 88,710/-

IV. ***Cost of production*:**

A. **Recurring cost** = Rs. 41,370/-

B. **Depreciation cost** Rs. 45,200/- × 10% = Rs. 4,520/-

C. **Margin money** Rs. 92,570/- × 10% = Rs. 9,257/-

D. **Interest of bank loan** Rs. (92,570 – 9257)/-
= Rs. 83313/- × 12.5% = Rs. 10,414/-

Total (A + B + C + D) = Rs. 65,561/-

V. ***Net profit from first year*:**

A. **Total income:** = Rs. 88,710/-

B. **Total cost of production** = Rs. 65,561/-

Net profit** (A – B) = Rs. 23,149/-

**As more numbers of rabbits are kept for breeding for the next year, the net profit will also be increased in second year.

References and Further Reading

Adrian, J., Legrand, G. and Frangne, R. (1981). Dictionnaire de biochimie alimentaire et de nutrition. Technique et documentation, Paris.

American Rabbit Breeders Association. www.arba.net.

Anon. (2004). Seventh Livestock Census 2003. Department of animal husbandry and dairying, Ministry of Agriculture, Government of India, New Delhi, India, pp. 39-73.

Arrington, L.R. and Kelley, K.C. (1976). Domestic rabbit biology and production. University Press of Florida, Gainsville.

British Rabbit Council. www.thebrc.org.

Chakraborty, A., Biswas, S. and Goswamy, A. (1999). Rabbit farming. First Edition. Published by Kalyani Publishers Inc. Pvt. Ltd., New Delhi.

Cheeke, P.R., Patton, N.M. and Templeton, G.S. (1982). Rabbit production. The Interstate Printers and Publishers Inc. Pvt. Ltd., Danville, Illinois.

Das, K.S., Chandrahas and Handa, M.C. (2003). Practical Manual on Rabbit Production Management (Course No.- LPM-313 under VCI curriculum), COVS, GADVASU, Ludhiana, pp. 1–57.

Dickerson, G.E. (1978). Animal size and efficiency: Basic concepts. *Animal Production*, 27: 367–379.

Food and Agricultural Organozation (FAO) (2010). Production Yearbook. Italy, Rome.

Hammond, J. and Marshall, F.M.A. (1925). Reproduction in the rabit. Oliver and Boyd, Edinburgh.

http://www.indg.in/agriculture/on-and-off-farm-enterprises/agri-onnoff-rabbit

http://www.jphpk.gov.my/English/rabbits.htm.

http://wildlife1.wildlifeinformation.org/s/0mlagomorph/leporidae/oryctolagus/oryctolagus_cuniculus/Oryctolagus_cuniculus_domesticus.html.

INRA. (1989). L'alimentation des animaux monogastriques: pore, lapin, volailles, Second edition, INRA, Paris, p. 282.

Lang, J. (1981). The nutrition pf the commercial rabbit, part 1, physiology, digestibility and nutrient requirements; part 2, feeding and general aspects of nutrition. *Nutrition Abstract Review*, Series B, Livestock Feeds, 51, pp. 197--25, 287–302.

Langer, P. (2002). The digestive tract and life history of small mammals. *Mammal Review*, 32: 107–131.

Lebas, F. (1971). Composition chimique du lait de lapine. Evolution au cours de la traite et en function du stage de lactation. *Ann. Zootech.*, 20: 185–191.

Lebas, F. (1975). Le lapin de chair, ses besoins nutitionnels et son alimentation pratique. Revised and amplified, ITAVI, Paris.

Lebas, F. (1981). Unpublished data.

Lebas, F. and Matheron, G. (1982). Livestock production in Europe perspectives and prospects. VIII. Rabbits. *Livestock Production Science,* 9: 235–250.

Lebas, F. and Colin, M. (1992). World rabbit production and research: situation in 1992. *Fifth World Rabbit Congress*. Vol. A: 29–54.

Lebas, F., Coudert, P., Rouvier, R. and Rochambeau, H. (1986). The rabbit- husbandry, health and production. FAO Publication, Rome.

Maertens, L., Janssen, W.M.M., Steeland, E., Wolters, D.F., Branje, H.E.B. and Jager, F. (1990). Tables de composition, de digestibilite et de valeur eneretique des matieres premieres pour lapins. Fifth Journees de la recherche cunicole, 12 to 13 December, Communication no. 577, ITAVI, Paris.

Mahajan, J.M. (1998). Angora rabbit. ICAR Publication, New Delhi.

Newton, R. and Susanaa, P. (1990). A manual for small scale rabbit production. FAO of United Nations.

Poole, T.B. (1975). The UFAW Handbook on the care and management of laboratory animal.

Proto, V. (1980). Alimentazione del coniglio da carne. *Coniglicoltura*, 17(7): 17–32.

Singh, U., Risam, K.S., Kumar, A. and Bhasin, V. (2004). Angora rabbit for wool production in India. *In*: National seminar on Angora rabbit wool and Cashmere production and utilization, 25-26th September, Manali, H.P., India, Souvenir of Lead Papers and Abstracts, pp-57-63.

Weisbroth, S.H., Flatt, R.E. and Kraus, A.E. (1974). The biology of the laboratory rabbit. Academic Press Inc. Pvt. Ltd., New York.

Index

D

E

F

N

O

P

R

W

Z